PLACING THE BORDER IN EVERYDAY LIFE

T0186732

Border Regions Series

Series Editor: Doris Wastl-Walter, University of Bern, Switzerland

In recent years, borders have taken on an immense significance. Throughout the world they have shifted, been constructed and dismantled, and become physical barriers between socio-political ideologies. They may separate societies with very different cultures, histories, national identities or economic power, or divide people of the same ethnic or cultural identity.

As manifestations of some of the world's key political, economic, societal and cultural issues, borders and border regions have received much academic attention over the past decade. This valuable series publishes high quality research monographs and edited comparative volumes that deal with all aspects of border regions, both empirically and theoretically. It will appeal to scholars interested in border regions and geopolitical issues across the whole range of social sciences.

Placing the Border in Everyday Life

Edited by

REECE JONES
University of Hawai'i at Manoa, USA

COREY JOHNSON
University of North Carolina at Greensboro, USA

Routledge
Taylor & Francis Group

LONDON AND NEW YORK

First published 2014 by Ashgate Publishing

Published 2016 by Routledge
2 Park Square, Milton Park, Abingdon, Oxfordshire OX14 4RN
711 Third Avenue, New York, NY 10017, USA

First issued in paperback 2016

Routledge is an imprint of the Taylor & Francis Group, an informa business

British Library Cataloguing in Publication Data
A catalogue record for this book is available from the British Library

The Library of Congress has cataloged the printed edition as follows:
Jones, Reece.
  Placing the border in everyday life / by Reece Jones and Corey Johnson.
      pages cm. – (Border regions series)
  Includes bibliographical references and index.
  ISBN 978-1-4724-2454-9 (hardback) – ISBN 978-1-4724-2455-6 (ebook) – ISBN 978-1-4724-2456-3 (epub) 1. Boundaries–Social aspects. 2. Borderlands–Social aspects. 3. Border security. I. Title.
  JC323.J66 2014
  320.1'2–dc23
                                                                                          2013049240

ISBN 13: 978-1-138-21863-5 (pbk)
ISBN 13: 978-1-4724-2454-9 (hbk)

Contents

SECTION III BORDER WORK BY NON-TRADITIONAL ACTORS AWAY FROM THE BORDER

List of Figures

List of Tables

Notes on Contributors

Anne-Laure Amilhat Szary (PhD University of Toulouse and Ecole Normale Supérieure) is a professor at Université Joseph Fourier, Grenoble, France and a researcher at the PACTE-CNRS unit. Her latest research concerns the interrelations between art and culture, analyzed through various projects on artists' intervention on borders and cultural production in and about contested places. She is the author of "Après la frontière, avec les frontières: dynamiques transfrontalières en Europe," editions de l'Aube, La Tour d'Aigues, co-edited with M.-C. Fourny, as well as of over 20 papers and other edited books and special issues of international journals.

Mathew Coleman (PhD University of California, Los Angeles 2005) is Associate Professor of Geography at The Ohio State University. He has research and teaching interests in political and legal geography, with a special emphasis on biopolitics and the politics of security. He has published in leading journals, such as *Antipode, Political Geography, Annals of the Association of American Geographers, Environment and Planning D: Society and Space, Geopolitics*, and *Law & Policy*.

Anthony Cooper (PhD Royal Holloway, University of London 2012) teaches in the department of Politics and International Relations at Royal Holloway University of London. His research interests coalesce around the theoretical and multidisciplinary study of borders and processes of bordering. He has published in leading journals on various aspects of border studies and associated subjects and is currently writing a book-length research monograph titled "Rethinking Borders: Globalization, Bordering, Connectivity," due to be published in 2014.

Corey Johnson (PhD University of Oregon 2008) is Associate Professor of Geography at the University of North Carolina at Greensboro. His research interests include borders, geopolitics of energy, and regional development policy, and his regional specialty is Central and Eastern Europe. He has published in leading journals such as *The Annals of the Association of the American Geographers, Political Geography, Geopolitics*, and *European Urban and Regional Studies*.

Reece Jones (PhD University of Wisconsin, Madison 2008) is Associate Professor and Chair of Graduate Studies in the Department of Geography at the University of Hawai'i at Mānoa. His book "Border Walls: Security and the War on Terror in the United States, India and Israel" won the 2013 Julian Minghi Outstanding Research Award from the Association of American Geographers. He has published research articles in the journals *Environment and Planning D: Society and Space,*

The Annals of the Association of American Geographers, Political Geography, and *Transactions of the Institute of British Geographers,* 2014.

Yakubu Joseph is a PhD candidate and Research Assistant at the University of Tübingen (Germany) and Research Coordinator of the International Institute for Religious Freedom (Bonn, Cape Town, Colombo). He holds a BSc in Sociology and Anthropology, an MSc in Sociology, and an MA in International Peace Studies. In his PhD dissertation, Yakubu Joseph focuses on the subject of ethno-religious conflicts, decentralization, and the national unity conundrum in Nigeria.

Vanessa Lamb is a doctoral candidate in the graduate program in Geography at York University, Toronto. Vanessa's research focuses on issues of nature, conservation, and competing claims over and for water/rivers, particularly in the context of Southeast Asia. Prior to commencing her PhD at York University, Vanessa completed her Masters at the University of Wisconsin-Madison and has also worked for Towards Ecological Recovery and Regional Alliance (TERRA), a Thai-based non-governmental organization.

Kenneth D. Madsen (PhD Arizona State University 2005) is an assistant professor of Geography on the Newark campus of The Ohio State University. His research focuses on the US–Mexico border with a particular emphasis on indigenous perspectives, border barrier structures, and tensions between interest groups at different scales. He has published in *American Indian Culture and Research Journal, Cultural Geographies, Geopolitics, Human Ecology, The Journal of Borderlands Studies,* and *Territory, Politics, Governance.*

Judith Miggelbrink (PhD University of Leipzig) is head of the research group "The production of space: state and society" at the Leibniz Institute for Regional Geography in Leipzig, Germany. She has led and carried out several research projects concerning the role of spatiality and territoriality in society. She is author of many publications, including a recent co-authored piece in *International Journal of Sociology and Social Policy* and *Nomadic and Indigenous Spaces. Productions and Cognitions* (2013), co-edited with Joachim Otto Habeck, Peter Koch, and Nuccio Mazzullo.

Emma S. Norman (PhD University of British Columbia, 2009) is a political and environmental geographer who has published widely in the areas of water governance, political ecology, and transboundary environmental justice. She is currently an assistant professor of Geography at Michigan Technological University with the Environmental and Energy Policy Program and Great Lakes Research Center. Her recent books include: *Waters without Borders? Canada, the U.S. and Shared Waters* co-edited with Alice Cohen and Karen Bakker (University of Toronto Press 2013) and *Governing Transboundary Waters: Canada, the*

United States and Indigenous Communities (Routledge: Earthscan series in Water Resource Management).

Chris Perkins (PhD in International Relations from Royal Holloway, University of London) is Lecturer in Japanese Studies at the University of Edinburgh. His research interests include discursive institutional approaches to borders, nationalism, and the sociology of contemporary Japan. Among his publications, he is co-author, with Anthony Cooper, of "Borders and status-functions: An institutional approach to the study of borders" in *European Journal of Social Theory* (2012) and co-author, with Chris Rumford, of the recently published "The Politics of (Un)fixity and the Vernacularisation of Borders" in *Global Society* (2013).

Rainer Rothfuss is Professor of Human Geography and Chair of the Examination Board at the Institute of Geography of the University of Tübingen. He leads a research team dealing with the issues of ethno-religious conflicts and development from the perspective of political geography. Other fields of research include participative planning and electric mobility as elements of sustainable urban development.

Chris Rumford is Professor of Political Sociology and Global Politics at Royal Holloway, University of London. Among his publications is *Cosmopolitan Spaces: Europe, Globalization, Theory* (Routledge 2008), which won the Gold Award in the Association of Borderland Studies' Past Presidents' Book Award 2010.

Angela Stuesse (PhD University of Texas, Austin 2008) is Assistant Professor of Anthropology at the University of South Florida. Her research and teaching interests include neoliberal globalization, migration, race, human rights, and methodologies of activist research. Her current work investigates the intensification of immigrant policing in Atlanta, Georgia. She has published in the journals *American Anthropologist, City and Society, Latino Studies, Southern Spaces,* and *Human Organization,* among others. Her book manuscript, *Globalization "Southern Style": Immigration, Race, and Work in the Rural US South,* explores how new Latino migration into Mississippi's poultry industry has impacted communities and prospects for worker organizing.

Jennifer Turner (PhD Aberystwyth University, 2014) is a post-doctoral researcher in the Department of Criminology at the University of Leicester. Trained as a geographer, her research focuses upon the emergence and transformation of the cultural-political formations associated with the penal system in the twenty-first century; concentrating more implicitly on the personal relationships that exist between offenders on the inside and the communities on the outside. She has authored papers for *Aether: The Media Geography Journal, Area, Geography Compass,* and *Space and Polity.*

Acknowledgments

Finishing an edited volume would be impossible without the assistance of dozens of people. We want to thank our editor at Ashgate, Katy Crossan, the Border Regions Series editor Doris Wastl-Walter, and the rest of the Ashgate editorial team for supporting the project and helping with the publication process. Jill Williams read and provided valuable comments on the introduction to the book.

Reece Jones would like to thank Sivylay, Rasmey, and Kiran for supporting his work and putting up with his many trips to the field for research. He thanks Mat Coleman, Joe Heyman, Corey Johnson, Joshua Kurz, and Joe Nevins for reading and commenting on earlier drafts of the paper. Julius Paolo created the map. Chapter 10 was originally published in the journal *ACME: An International E-Journal for Critical Geographers* 2014, 13(2). It is used with permission.

Corey Johnson thanks Apostol Apostolov and family, Christian Geiselmann, and Prof. Ulrich Ermann for valuable assistance in Bulgaria. He also gratefully acknowledges Candace Bernard and Robert Glickman for providing a dean's professorship that enabled him to work on this project during the 2012–13 academic year.

Kenneth Madsen would like to thank the many Tohono O'odham individuals, families, and elected officials who shared their experiences and insights with him over the past decade and a half. Their welcoming and ongoing support contributed to a more complete understanding of bordering processes in southern Arizona and northern Sonora. Special thanks is extended to Vivian Halton-Hedrington for providing detailed feedback on this manuscript.

Mat Coleman and Angela Stuesse received support for their research in Atlanta from the National Science Foundation and the University of South Florida Office of Research and Innovation. Graduate students Nolan Kline (USF) and Austin Kocher (OSU) provided essential research assistance.

Yakubu Joseph and Rainer Rothfuss acknowledge the support of the University of Tübingen to their Working Group on Human Geography, which has made it possible to undertake the field work and obtain relevant literature for this study. They are also thankful to many respondents and key informants in Jos during the field work for their cooperation and the invaluable insights gained from them about the phenomena explored. They salute the courage and desire of men and women of goodwill who are trying to build bridges of understanding and hold hands across the conflict divide.

Anne-Laure Amilhat Szary would like to thank her friend Professor Adriana Dorfman, distinguished Brazilian border geographer, who helped her a lot, first with the documentation of this text, then with her critical eye on it. It is dedicated to Maira and Elflay. Research was undertaken with the funding of the FP7 EUBORDERSCAPES project (FP7-SSH-2011-1, 290775).

Emma S. Norman would like to thank the participants of the Shared Waters Alliance for generously allowing her to attend meetings and to discuss their work. Thanks also to Eric Leinberger, cartographer at the University of British Columbia, for preparing the map found in her chapter. Thanks also to Karen Bakker and Matthew Evenden for commenting on earlier versions of this chapter. Lastly, Emma would like to thank her family—Chad, Parker, Luke—for their ongoing support.

Chapter 1

Where is the Border?

Corey Johnson and Reece Jones

The busiest land crossing into the European Union is at a remote place called Kapitan Andreevo, near the tripoint where Bulgaria, Turkey, and Greece meet. It is a typical hectic border crossing scene, a place where dozens of trucks and a steady lineup of cars and people on foot back up on either side of the border. The trucks, loaded with Turkish tomatoes and consumer goods, and cars full of tourists and migrants, await the mundane, technocratic processing so familiar to contemporary border crossers.

The region around Kapitan Andreevo is a good place to study the past and present of borders. At 100-years-old, this border is young by European standards, but makes up for its relative youth in its prominent position in the annals of regional and national folklore. Unnoticed by most border crossers is a statue of Captain Andreev—the namesake for the crossing—who was killed during the First Balkan War (1912–13) and through his martyrdom helped to secure a victory for Bulgaria over the mighty Ottoman Empire. Nearby towns and villages embody Billig's (1995) "banal nationalism" with public art celebrating border guards' service to the nation. In a 1980 census, one county had 13 stone statues depicting border guards, with six in a single village (Scarboro 2007).

Bulgaria is not alone in mythologizing its border agents. States around the world elevate the bravery of those who protect the Motherland from unwanted incursions, which keeps the home pure and protects the sanctity of sovereign borders between states. Romanticized statues of armed border guards, often accompanied by a canine companion, dot the borderscapes in many parts of the world, and serve as reminders of the vaunted place in national mythologizing reserved for these agents in the past.

During the Cold War in Europe, the tragic human stories of the Berlin Wall took their toll on this mythology—the heroes became not the lone sentry with his trusty dog, but the illicit border crossers themselves. The memory of these heroes is enshrined in countless indelible black and white photographs of escapees as they sought freedom on the other side of *die Todesstreife* (the death strip). All too often, of course, their treks ended in tragedy at the hands of border guards, who were not perceived as protectors of the Motherland, but instead symbolized in the public consciousness the cruelty of restricted human movement. The Berlin Wall came to symbolize the violence and exclusion of the East German state and the images of its fall in 1989 signified the collapse of the entire system.

These anecdotes of historical bordering, in spite of their key differences, share two things in common. First, in each case the geographical location of the border was explicit. The border was materially marked and cartographically mapped. It was patrolled by agents of the state who sought to control or prevent movement across the line, and those agents' jurisdictions typically were limited to the zone immediately adjacent to the border. The human experience with borders coincided with an encounter with the borderline itself, or the "razor's edge," across which travel was difficult but the promise of unimpeded movement once on the other side an ample reward (Schofield and Blake 2002, Cox 2008). Second, for the vast majority of people the experience was atypical; border crossing was an exceptional, not an everyday, experience. Humans were less mobile over long distances than they are today, a fact that globalization has changed for a substantial percentage of the world's population. The state viewed the political border as the key site to protect its sovereignty and thus the primary place where entry and egress was controlled.

State borders were never as strictly patrolled as these romanticized stories suggest. In many places borders were only firmly under the control of the state at the crossing points and the vast stretches between them were unguarded or lightly patrolled. The US–Mexico border, for example, was established in 1848 but the Border Patrol was not created until 1924 (Hernández 2010). Even then it was a small and underfunded force tasked with patrolling a 1969-mile border through rough, arid terrain. As late as 1990, the US–Mexico border did not have any fencing and the Border Patrol had just over 3,000 agents (Jones 2012). In 2013, there are over 21,000 agents and 670 miles of fence. Despite the mythologizing of a past golden era of borders, it is nevertheless true that in previous eras the vast majority of border work happened at the borderline and was carried out by the officers of the state.

Contrast these historical experiences of the border with a third anecdote from November 2011 in the United States. In the town of Vance, Alabama, a traffic cop made a routine stop because the license plate on a vehicle was not visible. The vehicle was driven by Detlev Hager, a German Mercedes-Benz executive visiting a factory in the town. In addition to checking the vehicle's license plate, the officer was also obligated to check the immigration status of the driver who had a foreign accent. Hager did not have his passport with him and was arrested and taken to jail on an immigration violation. A few weeks later a Japanese Honda executive was similarly arrested at a temporary police checkpoint in Alabama.

These stories about the auto executives made the news because the detained individuals were part of the supposedly borderless world of global commerce, but their experiences demonstrate a substantive shift in who does border work and where they do it. Whereas encounters with an international border were once exceedingly rare, today they are for many a daily occurrence, and increasingly not at the actual borderline. Contemporary borders, and by consequence the study of them, are decentered; the act of bordering happens far away from the border itself, as well as beyond the traditional realm of securing territory from unwanted incursions. In addition, more actors are called upon—or see it as their calling—to be the agents of the state in making the border. Indeed, the Alabama law and traffic checkpoints are specifically designed to identify and detain undocumented Latino workers in the state. The border checkpoint, in essence, briefly materializes thousands of miles from the actual line on a road in rural Alabama and is manned not by a federal agent but a local police officer. The border does appear to be everywhere (Balibar 2004).

However, the contributions to this book demonstrate that Balibar's assertion is only half right. There is no doubt that border work is now done at many new sites and by many new people but the fact that the arrest of the auto executives was newsworthy also demonstrates that these new borders are not designed to ensnare everyone, everywhere. Instead, the arrests were remarkable specifically because there is an assumption that there should not be a border for these wealthy executives. The implication is that there indeed is, and should be, a border in Alabama for poor undocumented workers. In this volume, we delineate precisely where these new borders are implemented, who is doing the new border work, and who specifically they are meant to locate.

The Everyday of Bordering: Conceptual and Theoretical Background

This book delves into the more recent geographies of border work, and it suggests that the previous reality of bordering—although it was never as tidy as it was represented or remembered—has nevertheless been both geographically displaced and partially supplanted by a new, more variegated reality. For every dramatic, hair-raising scramble across a wall or fence, there are multiple, mundane encounters with the border and its agents. Contemporary border work is technocratic, bureaucratic, and political—anything but romantic (though still sometimes romanticized). Border work is directed not only at transgressions of borders in the narrow sense of preventing a human from violating territorial sovereignty by crossing a line at the margins of a state's sovereignty, but also at border crossing by particular undesirable categories of goods and services, specific types of information, certain classes of humans, and nature. Yet even if the actual site of the border is less in focus than it was in pre-war Bulgaria or 1960s Berlin, for most people borders have become a more, not less, prominent feature of everyday life. This is true not only for the labor migrant or refugee, but also for the tourist, prisoner, protester, or auto executive on the way to a factory. For border scholars, this presents a host of new methodological and conceptual challenges: not only where do we go to find the border at work, but how does the concept maintain any sense of coherence once it is decentered.

This book offers some insights on these questions, but we acknowledge that much border work is still done by agents of the state at the border. Fences, walls, and even the lonely stone marker on an otherwise featureless landscape continue to serve their function in demarcating sovereign states' territorial limits and impeding free movements, but globalization and other structural changes altered the dynamic in notable ways (Johnson 2009, Newman 2001). Since the end of the cold war, the dominant way of conceptualizing borders has been as a dialectical relationship between the breaking down of borders and their increasing permeability on the one hand, and the maintenance and even fortification of borders on the other (Jones 2009, Newman 2006). Initially, the end of the cold war and emergence of the process of globalization were thought to be creating an increasingly borderless world where the dominance of liberal democracies and advances in transportation and communication technologies would connect distant places economically, culturally, and politically (Fukuyama 1992). The fall of the Berlin Wall, the removal of border checkpoints within the EU, and the creation of regional trade zones such as NAFTA all appeared to herald the possibility of a borderless world (Ohmae 1995, 1990).

Many other scholars disputed the claim of a borderless world by demonstrating that globalization was producing uneven boundary effects in which some movements were facilitated while others were restricted (Newman 1999, Ó Tuathail 2000, Sparke 2006). These seemingly opposing processes are in fact intricately interrelated: Some borders are opened to the movement of goods, capital, and certain desirable classes of humans while other boundaries are erected specifically to restrict undesirable classes of humans ("illegal immigrants," terrorists, etc.) (Geddes 2003, Häkli 2007, Mountz 2004). Much like other changes brought about by globalization, the shifting nature of political borders have both global as well as local significance. It is no coincidence, for example, that the European Commission calls their action plan on protecting its borders a "Global Approach to Migration: Priority Actions in Africa and the Mediterranean." The US Homeland Security strategy, meanwhile, advocates an extraterritorial approach to border security by "pushing borders out" beyond US sovereign territory (Hobbing and Koslowski 2009). Other scholars have also identified an inward gaze as boundary work that reinforces the border—from creating identity documents to raids to arrest undocumented immigrants—increasingly occurs away from the border itself within a sovereign state's territory (Amoore 2006, Appadurai 2006, Coleman 2005, 2009).

In the past 10 years the stigma associated with fortifying political borders disappeared and at least 23 security barriers were initiated or expanded worldwide, or more than double the number that were built during the entire Cold War (Hassner and Wittenberg 2009, Jones 2012). These include well-known projects in the United States (Nevins 2010) and Israel (Weizman 2007) but also in countries as diverse as Botswana, Thailand, Saudi Arabia, and Uzbekistan. Many more borders were hardened through the deployment of new security practices from increased patrols to new surveillance systems. In 2013, we estimate approximately 20,000 km (12,400 miles) of the world's borders are now marked with walls or barriers (Foucher 2007). An additional 18,000 km are "hardened" but unfenced boundaries (Rosière and Jones 2012). Contrary to expectations at the end of the Cold War, the current era of globalization has resulted in the most intensive and extensive period of bordering in the history of the world. These examples demonstrate that the process of globalization is accompanied by the rise of governance strategies, which are designed to cope with the risks inherent to the increasing mobility of humans and goods by patrolling beyond and within the territorial boundaries of the state (cf. Beck 1999).

While border security at the line is an important aspect of the everyday of bordering, we argue that sovereign bordering must be understood in a much broader context. The geographies of borders are more expansive, and as a result the conceptual tools we use to understand them must be expanded. This book, therefore, builds on recent work in geography and allied disciplines that has examined realignments in the relationship between concepts of territory, borders, and sovereignty (Brown 2010, Elden 2010) and on governmentality and security (Walters 2006, Bigo 2011). With the immigration reform debate in the US still unresolved and comparable discussions occurring in countries across the globe, border security is a topic of pressing concern in policy and academic communities alike. At the heart of these discussions is not only a debate about whether strategies by states to restrict access to its territory through barriers and expanded security practices are effective, but also ethical and philosophical questions about how, where, why, and by whom it is being done.

Non-traditional Actors and Locations for Border Work

The book makes three central arguments. The first is to identify the rescaling and expansion of border enforcement away from a top-down model of agents at border lines to border work by multiple local actors within the state's territory. "The implosion of border enforcement," as Mat Coleman and Angela Stuesse term it in their contribution to this volume, makes it clear that, not only are many non-state actors involved in the process of defining the boundaries of the group, but in many countries the work of patrolling the border often happens at sites away from the line. This process is evident in the post-9/11 refrain of "if you see something, say something," which deputizes every citizen to be the eyes and ears of the state (Vaughan-Williams 2008). More formally, in the US, the federal government is training local police officials in immigration enforcement which gives them the authority to inquire about immigration status and check documents. Previously this was solely the duty of the federal government but these new "secure communities" agreements move this authority to other non-federal actors at a variety of scales. This volume will analyze how and why sovereign states are expanding the power to enforce immigration laws to local law enforcement officials and the consequences this has for the lives of people within state territory.

In addition to the expansion of state border enforcement practices to interior locations, there is a growing number of state, quasi-state, and non-state actors doing the work of border making and enforcement at a range of locations that were not previously associated with the border. We theorize a shift away from the state itself to the position that a range of non-traditional actors have an interest in making borders and enforcing restrictions on movement both at the border and within the territory of the state (Parker and Vaughan-Williams 2009). The underlying argument is that borders are not just lines on the ground, but rather that they form an important part of our political imaginary that is predicated on an idea that a territory exists, that states inhabit it, and that people are bounded by it. As such, these borders in our minds are a fiction that shapes our conception of the world, our place in it, and our relationship to other humans (Castells 2010). In so doing, this fiction becomes a very powerful fact for the large numbers of people who encounter borders in their multiple locations in their daily lives. At the core of this book is the contested and often contradictory relationship between the narratives and practices of bordering the nation-state—the creation of inside and outside—on the one hand, and the real-life physical encounters with bordering practices on the other.

When the border is seen this way, as not simply a state artifact but a potent symbol that is firmly in the hands of individuals and organizations engaged in group making, it opens up a whole new range of actors who do border work every day for a variety of purposes (Walters 2006). These non-traditional border workers often have divergent goals that result in what Anthony Cooper, Chris Perkins, and Chris Rumford call, in a contribution to this volume, "the vernacularization of borders." Examples of this type of border work abound: right-wing organizations that want to protect a particular version of the cultural identity of the state; corporations involved in funding border security that rely on a narrative of insecurity and threat; and, in the US, state level legislation such as Arizona's SB 1070 or Alabama's HB 56 that complicate the lives of undocumented workers and result in a much broader definition of who does not belong. In the case of both pieces of anti-immigration legislation, one individual—a law professor, consultant, and later the Secretary of State of Kansas named Kris Kobach—played a central role in drafting the laws, himself becoming a border worker from his home in Kansas, as far from the US border as one can be. Other acts of everyday bordering are evident in the controversies over the construction of mosques in the US both near the World Trade Center site in New York, but also in places like rural Tennessee where the site of a proposed Islamic community center was protested and repeatedly vandalized, or in "price tag attacks" by Jewish settlers directed at Muslim citizens in Israel. Border work by non-traditional actors has multiple dimensions and functions and is carried out by different people in particular moments. What these instances share is a notion of insecurity in daily life that is projected onto the border.

The third contribution of this volume is to consider the concomitant devolution of border work to local officials and citizens and the everyday reimagining of the border in social practices through a critical geographic perspective that draws on qualitative and ethnographic field work. These empirical chapters ground the theoretical insights of border work done at multiple sites and scales by multiple actors in an analysis that identifies who these individuals are but also why they are engaging in this border work. These chapters range from investigations of how producers of television shows create a version of the reality on the US–Mexico border to Native American resource management practices to border work by local village councils along the Thai–Burma border. The diverse field sites in Africa, Asia, Europe, and North and South America demonstrate the range of locations and diversity of individuals engaging in contemporary border work. Informed by the rich output of border and border-related research during the past decade, this volume is an important guidepost to understand the rationale for examining the everyday politics of the border.

Conclusion: The New Face of Border Work

In the villages near Kapitan Andreevo, locals barely take notice of the border traffic, even the occasional illicit border crosser from Afghanistan or North Africa who has chosen this route to find a better life in Europe. In Berlin, meanwhile, the memory of the serious and lethal border work of the East German border guards has been reduced in some places to Disney-like kitsch. Nostalgia-seeking tourists, who have arrived in Berlin after breezing across Europe's Schengenland scarcely noticing borders between states, pause briefly to pose for photos with border guards dressed in authentic costume at Checkpoint Charlie.

Both Kapitan Andreevo and the former Berlin Wall symbolize the modern, sovereign state "container space" of the post-seventeenth century, which was a departure from prior forms of state bordering practices that were often overlapping and discontinuous. The idea of the state as a container and the border as a line on the map clarified belonging, compelled loyalties in all aspects of quotidian life, and shaped economic, cultural, and political interactions (Tilly 1990, Rokkan 1974). Much of the existing academic work on borders, in our view, has tended to analyze border securitization by accepting the core tenets of exclusive sovereign territory in the modern state system. In much of this work sovereignty serves, in John Agnew's words, more as "a background assumption than as a central theme of analysis" (Agnew 2009: 26). By showing how and in what forms border security discourses reflect and inform practice in border spaces themselves, and by tracing the networks security discourses and practices traverse, we argue that recent developments are anything but sovereign states acting alone on behalf of security. Instead, these changes represent a complex interplay of actors through which sovereignty is contingent and mediated at multiple scales which require new ways of thinking about who borders and how they do it.

Borders are key sites where sovereignty and territory, on the one hand, and networks and flows, on the other, intersect: they are at one level the quintessential manifestation of sovereignty and a territorial world, yet their very transformation over the past decade has been brought about by the reorganization of human migration, capital flows, and trade at the global scale. We need to move beyond the simple binary of *methodological territorialism* (Scholte 2005) on the one hand, and a type of *methodological networkism* on the other. The most naïve post-Cold War narratives of a flat world and the death of geography have been largely put to rest, but literature under a range of guises still maintains that globalization deterritorializes socio-spatial relations; that states are in decline (Brown 2010); that a distance-less world is emerging along with possibilities for a cosmopolitan politics; and that the world is composed of positive linkages between nodes of economic and political power. Echoes of this end of the spectrum can be found in geography in the "relational turn" (Jones 2010); discussions of a "flat ontology" without scale (Marston, Jones, and Woodward 2005); and non-representational theory (Thrift 2008). At the other end of the spectrum, scholars have posited that a postmodern world is still deeply territorially linked, and not even particularly postmodern. Much of this literature is coming from a realist IR perspective (e.g. Deudney 2007), but to some degree recent examinations of "territory" as an analytical category within geography exhibit this as well (Elden 2009). As Paasi has pointed out, territory versus network should not really be an "either—or" question (Paasi 2009), and there are promising suggestions of how to hybridize these positions such as the TPSN framework (Territories, Places, Scales, and Networks) (Jessop, Brenner, and Jones 2008). The contributions to this book demonstrate that border security is a spatial–territorial strategy of coping with the upheavals of an increasingly globalized world and that it is but one of a series of tightly regulated, seclusionary spaces of security humans have created in response.

The book is arranged into three sections. The first section makes the theoretical case for how to locate the who and where of contemporary border work. Chapter 2 by Anthony Cooper, Chris Perkins, and Chris Rumford argues for understanding the border work by non-state actors away from the border as the vernacularization of borders. In Chapter 3, Mat Coleman and Angela Stuesse identify the connections between a geopolitical approach to border policing at the line and a biopolitical strategy of creating uncertainty in the daily lives of immigrants through the devolution of border enforcement to local law enforcement. These two trends, explained in detail above, come together to move the border to multiple sites in our everyday lives.

The second section of the book analyzes border work by non-traditional actors in border areas. In Chapter 4, Emma Norman describes the challenges of transboundary governance of natural resources along the US–Canada border where citizens and environmental activists tried to develop cross border associations to regulate water pollution. Chapter 5 takes us south to the US–Mexico border where Kenneth Madsen explains the challenges faced by the Tohono O'odham community, whose ancestral lands cross both sides of the border and produce an array of different borders between communities on both sides. In Chapter 6, Vanessa Lamb describes how local residents and regional officials talk about the Thai–Burma border as it is potentially transformed by a large dam project. Chapter 7, by Judith Miggelbrink, wraps up the second section by describing how local small-scale traders negotiate the changing border regimes after the expansion of the European Union in Eastern Europe. These chapters demonstrate the multitude of individuals who do work that makes and unmakes the meaning and material existence of the border in their lives at the borderline.

The third section of the book analyzes border work by non-traditional actors away from the borderline. In Chapter 8, Yakubo Joseph and Rainer Rothfuss look at how local practices create boundaries between different ethnic and religious groups both in people's minds and in the urban landscape of Jos, Nigeria. In Chapter 9, Reece Jones argues that the producers of the National Geographic television show "Border Wars" create an imaginary story of the border in their television studio that valorizes the Border Patrol while dehumanizing and criminalizing immigrants. In Chapter 10, Anne-Laure Amilhat Szary discusses how the artists Marina Camargo and E. Nassar reflect on the recent evolutions of Latin American borders to tackle the territorial contradictions between transnational flows and nationalist politics. Chapter 11, by Jennifer Turner, points to prisons as another site where the inside and outside of the border is created through social and political practices. These chapters illustrate that the work of creating the border is often done far away the line on the map and the border on the ground.

The terms "freedom" and "security" are central facets of the political discourse in many parts of the world—perhaps even the defining concepts of these times. It is at borders, in the liminal spaces between polities, cultures, economies, and natures, where these concepts find their fullest expression as states seek to cope with the flows and movements that characterize our increasingly global world, and where humans are most exposed to the realities of our persistently territorial world. This book contends that there are very real symbolic and material consequences to bordering which reify, reinforce, and rework the connections between a sovereign state, a particular people, and a clearly defined bounded territory. However, the border work of the state in the twenty-first century is just as often done by non-state officials and at multiple sites within and beyond the territory of the state. This book contributes to our understanding of these new bordering practices by asking where the border is materialized and who makes borders in everyday life.

References

Agnew, J.A. 2009. *Globalization and Sovereignty*. Lanham: Rowman & Littlefield Publishers.

Amoore, L. 2006. Biometric Borders: Governing Mobilities in the War on Terror. *Political Geography*, 25(3), 336–351.

Appadurai, A. 2006. *Fear of Small Numbers: An Essay on the Geography of Anger*. Durham: Duke University Press.

Beck, U. 1999. *World Risk Society*. Malden, MA: Polity Press.

Bigo, D. 2011. Freedom and Speed in Enlarged Borderzones. *The Contested Politics of Mobility: Borderzones and Irregularity*, edited by V. Squire. New York: Routledge.

Billig, M. 1995. *Banal Nationalism*. London: Sage.

Brown, W. 2010. *Walled States, Waning Sovereignty*. New York: Zone Books.

Castells, M. 2010. *The Power of Identity*. 2nd ed. Malden, MA: Wiley-Blackwell.

Coleman, M. 2005. U.S. Statecraft and the U.S.–Mexico Border as Security/ Economy Nexus. *Political Geography*, 24(2), 185–209.

Coleman, M. 2009. What Counts as the Politics and Practice of Security, and Where? Devolution and Immigrant Insecurity after 9/11. *Annals of the Association of American Geographers*, 99(5), 904–913.

Cox, M. 2008. The Razor's Edge: A Review of Contiguity in Conflict Studies and an Argument for Redefining Neighbors. *International Journal of Social Inquiry*, 1(2), 237–257.

Deudney, D. 2007. *Bounding Power: Republican Security Theory from the Polis to the Global Village*. Princeton: Princeton University Press.

Elden, S. 2009. *Terror and Territory: The Spatial Extent of Sovereignty*. Minneapolis: University of Minnesota Press.

Elden, S. 2010. Land, Terrain, Territory. *Progress in Human Geography*, 34(6), 799–817.

Foucher, M. 2007. *L'obsession des Frontières*. Paris: Perrin.

Fukuyama, F. 1992. *The End of History and the Last Man.* New York: Free Press.

Geddes, A. 2003. *The Politics of Migration and Immigration in Europe*. London; Thousand Oaks, CA: SAGE Publications.

Häkli, J. 2007. Biometric Identities. *Progress in Human Geography*, 31(2), 139–141.

Hassner, R. and J. Wittenberg. 2009. Barriers to Entry: Who Builds Fortified Boundaries and Are They Likely to Work? Paper to the annual meeting of the American Political Science Association. Toronto, 6 September.

Hobbing, P. and R. Koslowski. 2009. *The Tools Called to Support the "Delivery" of Freedom, Security, and Justice: A Comparison of Border Security Systems in the EU and in the US*. Brussels: European Parliament, Document PE 410.681, February.

Jessop, B., N. Brenner, and M. Jones. 2008. Theorizing Sociospatial Relations. *Environment and Planning D: Society and Space*, 26(3), 389–401.

Johnson, C.M. 2009. Cross-border Regions and Territorial Restructuring in Central Europe: Room for More Transboundary Space. *European Urban and Regional Studies*, 16(2), 177–191.

Jones, M. 2010. Limits to "Thinking Space Relationally." *International Journal of Law in Context*, 6(3), 243–255.

Jones, R. 2009. Geopolitical Boundary Narratives, the Global War on Terror and Border Fencing in India. *Transactions of the Institute of British Geographers*, 34(3), 290–304.

Jones, R. 2012. *Border Walls: Security and the War on Terror in the United States, India and Israel*. London: Zed Books.

Marston, S.A., J.P. Jones, and K. Woodward. 2005. Human Geography without Scale. *Transactions of the Institute of British Geographers*, 30(4), 416–432.

Mountz, A. 2004. Embodying the Nation-state: Canada's Response to Human Smuggling. *Political Geography*, 23(3), 323–345.

Nevins, J. 2010. *Operation Gatekeeper and Beyond: The War on Illegals and the Remaking of the U.S.–Mexico Boundary*. 2nd ed. New York: Routledge.

Newman, D. 2001. Boundaries, Borders, and Barriers: Changing Geographic Perspectives on Territorial Lines. *Identities, Borders, Orders: Rethinking International Relations Theory*, edited by M. Albert, D. Jacobson, and Y. Lapid. Minneapolis: University of Minnesota Press.

Newman, D. 2006. The Lines That Continue to Separate Us: Borders in our "Borderless" World. *Progress in Human Geography*, 30(2), 143–161.

Newman, D. ed. 1999. *Boundaries, Territory and Postmodernity*. London; Portland, OR: Frank Cass.

Ó Tuathail, G. 2000. Borderless Worlds? Problematizing Discourses of Deterritorialization. *Geopolitics at the End of the Twentieth Century: The Changing World Political Map*, edited by N. Kliot and D. Newman. London; Portland, OR: Frank Cass, 139–154.

Ohmae, K. 1990. *The Borderless World: Power and Strategy in the Interlinked Economy*. New York: HarperBusiness.

Ohmae, K. 1995. *The End of the Nation State: The Rise of Regional Economies*. London: HarperCollins.

Paasi, A. 2009. Bounded Spaces in a "Borderless World": Border Studies, Power and the Anatomy of Territory. *Journal of Power*, 2(2), 213–234.

Parker, N. and N. Vaughan-Williams. 2009. Lines in the Sand? Towards an Agenda for Critical Border Studies. *Geopolitics*, 14(3), 582–587.

Rokkan, S. 1974. Entries, Voices, Exits: Towards a Possible Generalization of the Hirschman model. *Social Science Information*, 13(1), 39–53.

Rosière, S. and R. Jones. 2012. Teichopolitics: Re-considering Globalisation through the Role of Walls and Fences. *Geopolitics*, 17(1), 217–234.

Scarboro, C.A. 2007. *Living Socialism: The Bulgarian Socialist Humanist Experiment*. Doctoral Dissertation, History, University of Illinois, Urbana-Champaign.

Schofield, C.H. and G.H. Blake. 2002. *The Razor's Edge: International Boundaries and Political Geography: Essays in Honour of Professor Gerald Blake*. London; New York: Kluwer Law International.

Scholte, J.A. 2005. *Globalization: A Critical Introduction*. 2nd ed. New York: Palgrave Macmillan.

Sparke, M.B. 2006. A Neoliberal Nexus: Economy, Security and the Biopolitics of Citizenship on the Border. *Political Geography*, 25(2), 151–180.

Thrift, N.J. 2008. *Non-representational Theory: Space, Politics, Affect*. London: Routledge.

Tilly, C. 1990. *Coercion, Capital, and European States, AD 990–1990*. Cambridge, MA: Blackwell.

Vaughan-Williams, N. 2008. Borderwork Beyond Inside/Outside? Frontex, the Citizen-detective and the War on Terror. *Space and Polity*, 12(1), 63–79.

Walters, W. 2006. Border/Control. *European Journal of Social Theory*, 9(2), 187–203.

Weizman, E. 2007. *Hollow Land: Israel's Architecture of Occupation*. London: Verso.

SECTION I
Theorizing the Border in Everyday Life

Chapter 2

The Vernacularization of Borders

Anthony Cooper, Chris Perkins, and Chris Rumford

Introduction

A major shift in border studies in recent years has been away from an exclusive and primary concern with conventional nation-state borders (the external edges of a polity) to a concern with borders being dispersed throughout society and found "wherever the movement of information, people and things is happening and is controlled" (Balibar 2004: 1). This move towards the study of "diffused" borders has been partially offset by the post-9/11 preoccupation with securitization and surveillance. Rosière and Jones (2012) have noted the "hardening" of diffused borders through the construction of walls or fences, a process they term "teichopolitics." The resultant and somewhat contradictory dynamics of contemporary border studies are summed up by Lyon (2013: 6) in the following terms, "Even national borders, which once had geographical locations—however arbitrary—now appear in airports distant from the 'edge' of the territory and, more significantly, in databases that may not even be 'in' the country in question." The "biometric border" (Amoore 2006) is emblematic of these major shifts, as is the idea of "remote control" (obliging airlines to conduct their own security checks on passengers and their travel documents) (Guiraudon and Lahav 2000). The study of borders which no longer only take the form of securitized perimeters, and which are sometimes enacted by commercial agents, and can possibly be controlled from a distant center suggests a radically transformed landscape of borders and bordering. However, certain key aspects of borders (and the way we study them) remain largely untouched by much current thinking, including a continued preoccupation with the borders of the state. Similarly, the role of citizens in bordering activity remains largely the same in the border studies imagination despite the ability of "remote control" to recruit a range of actors to carry out work on behalf of the state.

There is evidence of dissatisfaction with the state/security/mobility agenda which continues to dominate the field of border studies. The need to identify new meanings of the border not tied to the state (Bauder 2011) has emerged as a key theme in the critical literature, as has its corollary, the inadequacy of conceptualizations of the border (Rovisco 2010). In this chapter we seek to advance this emerging agenda by shifting the focus in a particular direction, one that acknowledges both the role of ordinary people in making, shifting, and removing borders—the idea of "borderwork" (Rumford 2007, 2008, 2012, forthcoming) and the variety of roles that borders can fulfill in addition to being markers of (state) division. A border studies which embraces the vernacularization of borders allows for a shift of emphasis from state bordering, securitization, and the regulation of mobilities to a concern both with the role of borders in "the politics of everyday fear" (Massumi 1993) and bordering as a political resource for citizens who are able to both contest nation-state bordering practices and institute their own versions of borders. Borders can be political resources in the sense that they can be drawn upon by a range of actors who seek to either selectively regulate mobility, use the border as a staging post which connects to the wider world, or simply use the border as a way of navigating the multiplicity of spaces which characterize a world in perpetual motion.

Arguably the single most important conceptual development laying the ground for the emergence of a vernacularized study of borders is Balibar's insight that borders are increasingly diffused. To sit alongside this we propose a second key innovation: the idea of borders as "engines of connectivity." Borders not only divide; they also connect, both to the other side of the border and also, on occasions, far beyond (see Cooper, forthcoming). To better understand how borders can be utilized for connectivity, at various sites, by various actors in their everyday practices we need to engage with the idea of "scale" in relation to borders, drawing upon van Schendel's (2005) insight that borders make it possible for those at the border to "jump scales," thereby making possible global connectivity. We develop this idea at length later in the chapter. Before turning our attention to scales and connectivity we must look at one other key factor in the emergence of a vernacularized border studies, without which the conceptual advances represented by "diffused" borders and "scale jumping" would have not found much purchase. This is the recognition that borders have changed in significant ways in a "world in motion" and that what is needed is a conceptual toolbox for apprehending this changed landscape of borders and bordering. A good indicator that borders are changing is to be found in the range of approaches that now exist in contemporary borders cartography. Representing borderlands, diffuse borders and networked borders, for example, in conventional cartographic terms is problematic (Walters 2009, Gschrey 2011, Kopper 2012). Nevertheless, some very interesting work on alternative border mapping has emerged (Multiplicity 2005, Casas-Cortes and Cobarrubias 2010, Kramsch 2011). For example, Multiplicity's (2005) mapping of border flows and categorization of the border (as funnels, pipes, folds, sponges, enclosures, and "phantom limbs") creates a visually arresting depiction of human traffic in border zones. It is an attempt to represent the changing nature of borders through non-conventional mapping techniques.

The Changing Nature of Borders

It is important to outline the key changes in the nature of borders that have emerged in the contemporary literature and the debates contained therein. The first change is the idea already mentioned above that "borders are everywhere." Importantly, this is more than the recognition that multiple sites of bordering now exist: at airports, Eurostar terminals, and maritime ports. It suggests that borders exist at other locations, many of which would not be thought of as borders in the conventional sense: in travel agencies and other offices where travel documents are issued and databases checked, along motorways where trucks are scanned and car number plates monitored (Walters 2006), and on the internet where credit card shopping makes possible the "transaction mining" of information for security purposes (Amoore and de Goede 2008). The border can even be said to exist at private airfields and on farms where the security of small aircraft and supplies of fertilizer is an anti-terrorist priority. In respect of the latter, in the UK the National Counter Terrorism Security Office (NaCTSO) has urged farmers to secure inorganic nitrogenous fertilizers against the possibility of appropriation by terrorists (www.secureyourfertiliser.gov.uk).

The second change is the recognition that borders mean different things to different people and act differently on different groups; borders are designed to separate and filter. Taking up this theme in a recently published reference volume, Wastl-Walter (2012) writes that borders:

> are manifested in diverse ways, and have various functions and roles. They can be material or non-material and may appear in the form of a barbed-wire fence, a brick wall, a door, a heavily armed border guard or as symbolic boundaries ... while a brick wall may represent security for some, for others, it may be a symbol of suppression.

It could be argued that Wastl-Walter does not go far enough in this statement. There are many kinds of borders that exist between brick walls and "symbolic boundaries," such as e-borders and the "juxtaposed" borders found along the Eurostar route, not to mention the non-state borders that we argue are increasingly important. Nevertheless, what is particularly interesting about Wastl-Walter's formulation is the recognition that borders can mean different things to different people: security or suppression, walls or bridges, barriers or turnstiles. This change is also captured by Balibar's (2002) idea of polysemy which suggests that borders are becoming "asymmetrical membranes" (Hedetoft 2003) or acting like "firewalls" (Walters 2006). All of these metaphors point to borders being designed to allow the passage of "desirables" while keeping out "undesirables." The UK has developed polysemic borders in an attempt to create "security in a global hub" (Cabinet Office 2007) through e–borders designed to be "open to business but closed to terrorists and traffickers." The border is polysemic precisely because it works very differently on those who have "trusted traveler" status compared to those upon whom suspicion falls on (or before) entry, e.g. those travelling on a student visa, or those without adequate documentation.

The third change is further recognition that the location of borders is changing, away from the edges of a nation-state. It was a key element in a recent "state of the debate" report published by the European Commission: "the study of borders has moved away from an almost exclusive concern with the borders between States in the international system, to the study of borders at diverse socio-spatial and geographical scales, ranging from the local and the municipal, to the global, regional and supra-State ..." (Kolossov 2012: 3). In fact, borders can now be remote and distant from the territory they are designed to protect. The UK is now developing "offshore borders all over the world"[1] in order to prevent undesirables from starting their journey to the UK. The Eurostar train link has introduced "juxtaposed" borders so that UK passport control takes place at Gard du Nord and French passport control at St Pancras. Paraphrasing Lahav and Guiraudon's (2000) we can say that "borders are not always at the border." The fourth change follows logically from the first three: borders are increasingly becoming mechanisms to "control mobility rather than territory" (Dürrschmidt and Taylor 2007: 56). The traditional idea that borders lock down territory or form a security perimeter for the sovereign nation-state is supplemented with the idea of the border as a manageable conduit, speeding up transit where necessary, blocking passage when required.

Dimensions of Vernacular Borders

Having established the context for the emergence of, and need for, vernacularized border studies let us now turn our attention to the research agenda to which it gives rise. After outlining the three key dimensions of vernacularized border studies the focus in subsequent sections will be on one of them; "bordering as connectivity," the aim being to better establish the potential of this particular perspective. The first dimension stems from the idea that bordering is no longer only the business of the nation-state, which should not be taken to imply that only bottom-up bordering is of interest. In Europe, the EU has emerged as a major actor in the business of creating, relocating, and dismissing borders. The EU shifts the borders of Europe every time it enlarges, it turns national borders into European borders, it regulates and harmonizes European borders through Frontex, its borders agency, and it has the power to decide where the important borders in Europe are to be found (Rumford 2006). Elsewhere, one of us has argued that this is evidence that Europe possesses "cosmopolitan borders" (Rumford 2007). It could be argued that EU bordering is state bordering by a different name, and, as such, the EU's ability to re-border Europe does not advance the vernacularization thesis. But as Papadopoulos et al. (2008) point out, Europe's borders are constructed wherever they are needed by the EU, not according to nation-state preferences. In any case, the key element here is that bordering activity in Europe is increasingly conducted both above the level of the state and below it. From the vernacularization perspective it is more important to consider the activity of ordinary people in contributing to processes

---

1   Home Office Minister, Liam Byrne, quoted in Home Office News Release (106/2007) issued by The Government News Network on 18 June 2007.

of bordering, introduced here under the rubric of "borderwork." Good examples of this phenomenon can be found in the UK: at Berwick-upon-Tweed, on the English–Scottish border, where residents have sought to redraw the Scottish border to include their town, and in a number of English towns (e.g. Totnes, Brixton, Lewes, Bristol) where business people have championed the creation of local currencies so as to stop money leaching away from the local economy (as part of the wider Transition Towns initiative). The introduction of this local currency effectively draws a new boundary around the town designed to (symbolically) regulate capital flows and build confidence in the local economy. This dimension of citizen bordering is largely absent from the regular border studies literature in which it is understood that people can utilize borders for their own advantage, as smugglers, tourists, and market traders frequently do, and that people are active in marking the state border through shows of nationalist fervor or grass-roots protest, for example. However, what is rarely considered is that citizens (and indeed non-citizens) may be active in constructing or dismantling borders as a form of political opportunism or self-empowerment. This borderwork may or may not take place at the edge of a polity, but is in fact more likely to take the form of bordering dispersed throughout society, as Balibar has theorized.

The second dimension is the importance of including a perspective from the border. This builds upon existing work on the border as a potential site of cosmopolitan activity. For example, Walter Mignolo has argued that "border thinking" is a core component of critical cosmopolitanism. For Mignolo, critical cosmopolitanism comes from the "exterior of modernity," in other words coloniality (Mignolo 2000: 724). Border thinking—"the transformation of the hegemonic imaginary" from the perspective of the excluded—is a tool of critical cosmopolitanism (Mignolo 2000: 736–37). We can usefully extend this and propose that "seeing from the border" is a key dimension of the vernacularization of borders. "Seeing from the border" cannot be reduced to the idea that it is possible to view a border from both sides (Rumford 2011). More than "looking both ways" *across* a border we need to aspire to look *from* the border. As borders can be found "wherever selective controls are to be found" (Balibar 2002: 84–85), seeing like a border does not equate to "being on the outside and looking in" (or looking out from the watchtower to the wilderness beyond). As we have seen, borders are not necessarily always working in the service of the state. When seeing like a state one is committed to seeing borders as lines of securitized defense (Scott 1998). Borders do not always conform to this model. In a desire to shore up what may be perceived as the ineffectual borders of the nation-state, borderworkers may engage in local bordering activity designed to enhance status or regulate mobility: gated communities, respect zones, resilient communities of CCTV watching citizens. These borders are not necessarily designed to enhance national security.

The third dimension is borders as connective tissue, an idea, as we have already seen, which builds upon van Schendel's (2005) notion that borderlanders are able to "jump" scales (local, national, regional, global) through their everyday practices and therefore do not experience the national border only as a limit: what forms a barrier to some can present itself as a gateway to others. People can "invoke" the scale of the border themselves: as a "local" phenomenon, a nation-state "edge," or as a transnational staging post, thereby allowing them to experience the border as a conduit. This insight means that we can take issue with the more mainstream idea, expressed for example by Häkli and Kaplan (2002: 7), that "cross-border interactions are more likely to occur when the 'other side' is easily accessible, in contrast to when people live farther away from the border." People can not only possess an interest in distant borders but have realistic expectations of having a "presence" there. Consider, for example, the locals in an Australian bar who spend time online monitoring the US–Mexico border via live webcam links. In fact the US–Mexico border can now be policed by anyone with an Internet connection. "Once logged in the volunteers spend hours studying the landscape and are encouraged to email authorities when they see anyone on foot, in vehicles or aboard boats heading towards US territory from Mexico" (Luscombe 2009). In this way, borders can work to provide transnational or global connectivity by allowing people to project themselves beyond their locality by constructing new networking opportunities.

Borders as Engines of Non-proximate Connectivity

We will now continue to look at connectivity as a key dimension of border vernacularization in much more detail. More specifically, in this section we have a particular interest in how border/ing can be utilized (or have the potential) to connect well beyond localities within which the border may be situated. As well as being a novel aspect of border vernacularization that merits detailed discussion in its own right, looking at borders through a lens of non-proximate connectivity is useful in the way it overlaps with the other two aspects detailed above. First, such connectivity does not simply involve traditional state borders; on the contrary, non-state borders, such as PGI (Protective Geographic Indication) designation, UNESCO World Heritage designation, or monument sites (see Cooper and Rumford 2013) can be used for connective ends just as much as state borders. Secondly, this bordering connectivity can take place well away from state edges and peripheries, providing a novel way in which important and tangible borders are dispersed throughout society. A final advantage of such a focus is that non-state actors produce and maintain this connectivity vis-à-vis borders. As discussed in the introduction, this offers a particular aspect of borderwork in the sense that ordinary people can use the connectivity potential afforded by particular borders as political resources.

Certain aspects of borders have often been discussed in terms of connectivity. Conceptually, they connect an inside to some outside (van Houtum, Kramsch and Zierhofer 2005), they can form "in-between" spaces of contact and negotiation (Martinez 1994, Boer 2006, Konrad and Nicol 2008) and, particularly in the context of securitization, have been described as "membranes" (Hedetoft 2003) that simultaneously channel (connect) some while blocking others. Conceptual geopolitical imagery and securitization aside, such connectivity relies on, or is the product of, proximate crossing and contact, often further framed in terms of the experiences of border crossers and borderland dwellers (see, for example, Strüver 2004). In this sense the border, as Konrad and Nicol (1998: 32) put it, becomes a "zone of interaction where people on one side of the border share values, beliefs feelings and expectations with people on the other side of the border." Although it is worth mentioning here that Martinez (1994: 2–4) would qualify this level of "openness" as an example of an "interdependent borderland," in which favorable international relations between two or more states conditions cross-border interdependence, rather than the less favorable "alienated" or "co-existent" borderlands within which cross-border interaction is severely limited if at all present. Yet as far as this chapter is concerned, outlining as it does a case for vernacular perspectives on borders, the connectivity briefly outlined thus far—the crossing, interactions, and encounters—tend to rely on what is immediately on either side of the border. Likewise there is a sense that the state (border) dominates and thus frames the type of connections taking place because these connections tend to be situated within the close proximity of a recognizable state border.

Looking at the relationship between borders and scale provides a good foundation upon which we can begin to understand *non-proximate* connectivity. It has, of course, been observed that borders are "multi-scalar," operating and transforming at different geographical levels and should be studied accordingly (Newman and Paasi 1998: 200). That said, common discussions concerning scale and borders more often than not refer to, or at the very least privilege, the state as the dominant frame in which to study borders (Kramsch 2010). When discussing scale, for example, Newman and Paasi (1998: 197) posit that while borders should be examined at all levels of scale it is the local sphere of everyday life "where the meanings of (state) boundaries are ultimately reproduced and contested."

Scale is also implicit in van Schendel's work on the Bangladeshi borderlands in which he acknowledges the importance of local everyday life but also departs from it, at least in the way Newman and Paasi describe it. Placing his work in context, van Schendel is keen to move away from previous/still dominant geopolitical imaginaries that posit a neat and easy correlation between state, territory, society, and nationhood. To this end, observing the spatiality of the border as a case in point, van Schendel (2005: 44) argues that "borders not only join what is different but also divide what is similar." Like Newman and Paasi, van Schendel (2005: 44) looks at the ways in which border meanings are contested and reproduced within the local sphere, but deviates from conventional interpretations in the sense that international borders are "crucial localities" for observing "how global restructuring affects territoriality." Unlike Newman and Paasi, therefore, borders must be "understood as dynamic sites of transnational reconfiguration" (ibid.), with the driving force being the ability for borderlanders to also "jump scales," that is, to scale beyond the state rather than simply constructing, reaffirming, and contesting it from a local perspective. For van Schendel's borderlanders, the state border is not an intermediary level between the local and the global, but rather represents both the local and the global giving borderlands the ability to be involved in transnational processes in their everyday lives. Van Schendel (2005: 49) sums this up thus:

> For borderlanders, the state scale is not overarching and does not encompass the more 'local' scales of community, family, the household or the body. On the contrary, to them it is the state that, in many ways, represents the local and the confining, seeking to restrict the spatiality of borderlanders' everyday relations. Scales that most heartlanders experience as neatly nested within the state scale—face-to-face relations of production, marketing networks, or community identities—are experienced very differently by borderlanders. In their case, these scales are often less 'local' than the state; they breech the confines of that scale, spill over its limits, escape its mediating pretensions, and therefore set the scene for a specific borderland politics of scale.

It is useful to point out here that, for van Schendel, scale politics involves the inability of the state scale to prevent clandestine (unauthorized) cross border activity. The state scale, in other words, is unable to achieve complete hegemony because it is constantly being challenged by the restructuring/rescaling capabilities of borderlanders—what van Schendel (2005: 55) has summed up as "everyday transnationality." A crucial component of this everyday transnationality is the ability to construct internal cognitive maps, whereby borderlanders can envisage and situate themselves across multiple scales of which the state is only one. Borders for van Schendel, therefore, are defined as being more than barriers or limits however much they are institutionalized by nation states. State borders represent only one important scale amongst others and, in this regard, the state border does not operate as an all-defining framework, which encompasses "lower" scales such as the "local" or the "community" and so on. Acting as a staging post, the (state) border can be utilized to project and connect identities and social relations outside of and beyond the immediate border.

Importantly for us, van Schendel's work begins to move away from a hierarchical model of scale, which is vital to discussing non-proximate connectivity in relation to borders. In this regard, Leitner (2004) argues that conventional approaches to observing scale—defined as level and size—fail to take into account the relations amongst different scales, as well as the ways in which processes supposedly operating at different scale levels influence each other. Leitner (ibid.) argues that scale politics is about "relations of power and authority by actors and institutions operating and situating themselves at different spatial scales." There is a tendency for the reproduction of hierarchical imaginaries of scale—small, large, and so on—to override more horizontal forms of social relations. Yet for Leitner (2004: 252), the different spatiality of networks ruptures "the familiar scaled political map," and in doing so they "connect to places horizontally across the bounded spaces of political territorial entities, which themselves are part of scalar state structures." This is similar to what Brenner (2004: 605) has observed as tangled scalar hierarchies:

> The meaning, function, history and dynamics of any one geographical scale can only be grasped relationally, in terms of upwards, downwards and sidewards links to other geographical scales situated within tangled scalar hierarchies and dispersed interscalar networks.

Importantly, we argue that borders have the capacity to "make horizontal" the hierarchical scale structure they are often implicit within and/or the product of. That is to say horizontal networks do not simply traverse borders—indicative of what Bude and Dürrschmidt (2010) term "flow speak"—that rather borders in and of themselves can function as "interfaces" that "level out" socio-spatial relations (Cooper and Rumford 2013). The border, when theorized in terms of the connection on offer here, reduces the verticality and compartmentalized nature of scale traditionally understood, making it tangled, warped, and more horizontal, and allowing contact with "Others" that would normally be hierarchically separated and distant (see also Cooper, forthcoming). By seeing from the border, non-proximate connectivity is brought to the fore, and we can begin to understand that what appears to be distant, unfamiliar, and "beyond" to some (a vertical imposition of scale), can be equally local, similar, and "near" to others (what could be described as a more horizontal imposition of scale).

This way of looking at borders further resonates with the idea of vernacularization because it also takes into account different types of border, constructed, maintained, and/or performed by different people. Our understanding of borders as engines of non-proximate connectivity, in other words, need not be limited to, or be under the exclusive remit of, recognizable borderlands and borderlanders. The residents of the English town of Berwick-upon-Tweed, for example, have shown a remarkable capacity to use the close proximity of the English–Scottish border to project their uniqueness—that is neither English nor Scottish—well beyond the locality of the border by forging networks and creating cultural events of global significance that would not be possible without the border. Likewise, but located well away from more traditional national borders/borderlands, the English town of Melton Mowbray has successfully established PGI (Protective Geographical Indication) designation (bestowed by the European Commission) that extends its identity well beyond the locality of the PGI border, projecting the town onto the "global food map" (see Cooper and Rumford 2011). In the next section, we will discuss a more detailed and nuanced example of border vernacularization taking place in Japan which shows how non-state borders are not only being created and dispersed, but also how "citizens" can move between institutional layers creating new connections because of the borderwork taking place.

Sakaguchi Kyōhei's Use of the Vernacularized Border

This final section offers a case study that illustrates the three different elements of the vernacularized border detailed above. Taking cues from the work of Amoore and Hall (2010, 2013), who have explored bordering practices, performance, and artistic interventions, we focus on the work of Japanese activist and performance artist Sakaguchi Kyōhei (2012), who set about constructing a new state as a critique of the Japanese government's response to Japan's March 2011 triple disaster. Sakaguchi's ad hoc state problematizes sovereignty, state territory, and established bordering practices, and although Sakaguchi does not explicitly talk about his work in terms of bordering, as we show in the following, he nevertheless draws attention to the performative, connective, and scalar aspects of borders. Sakaguchi's example also illustrates what Bauder (2011) refers to as "aspect-seeing" in relation to the border. Bauder (2011: 1132) argues that the concept of aspect-seeing draws attention to the different "uses of a border and border practices," and that "various aspects of the border represent meanings and material practices that cannot be unified into a stable and coherent concept." Sakaguchi himself plays on this concept of aspect seeing, which he refers to as seeing different layers (*reyā*) of action within public space. This forms an integral component of his political project, which seeks to destabilize the routinized articulation of the state and territorial sovereignty by drawing new borders both within and beyond the Japanese state.

Finally this example illustrates another component essential to understanding the conditions in which vernacularization of the border becomes possible, namely the importance of changes in what Cooper and Perkins (2012) call the "background assumptions" that naturalize particular institutional arrangements. In other words, a bordering process that appears natural may been thrown into a different light due to changes in the background conditions in which the bordering process developed, and it is these changes that draw attention to different aspects, meanings, and material practices of the border. Similarly, vernacularization of the border also relies on particular background conditions that enable actors to legitimize their borderwork. To draw on the example of Transition Towns discussed above, Perkins and Rumford (2013: 278) have argued that the 2008 financial crisis produced both discursive and material conditions that enabled residents of Stroud to set up an economic border in the form of the Stroud Pound. Advocates of this economic border justify it through appeals to generalized worries over capital flight, jobs, and the continued salience of local practices, while simultaneously linking their town to other Transition Towns running similar schemes both in the UK and abroad (ibid.).

Japan's 2011 triple disaster provides the context for Sakaguchi's borderwork. The earthquake, which struck on March 11, 2011, was the fourth biggest in recorded history and produced a tsunami that devastated the northeastern seaboard of Japan. This tsunami also breached the protective wall surrounding the Fukushima Daiichi nuclear plant, causing damage that led to a meltdown.

To date Japanese authorities have counted 15,854 dead from the earthquake and tsunami (BBC 2012), while the impact of the nuclear meltdown remains uncertain. Furthermore, the government's perceived weakness and poor policy response led to large protests across the country. In the weeks and months following the disaster, information regarding the structural links between the Japanese state and Tokyo Electric Power Company (TEPCO), the electricity company that built the Daiichi plant, emerged, and citizen's groups and "free" journalists have criticized the mainstream media for what was perceived as overly conservative coverage of the unfolding disaster and the protest movement. This event can be conceptualized as producing a crisis of state legitimacy which enabled various actors in Japan to "see" hitherto suppressed contradictions internal to the Japanese state, such as the conflict between Japan's experience of nuclear devastation and its championing of nuclear power in the postwar era (Shun'ya and Loh 2012). Sakaguchi has himself recognized the importance of 3/11 in shifting the context for debate in Japan, arguing that:

> If you talked about these things in the past people would talk about you as if you were crazy, but when something like 3.11 happens everything flips over … It's certainly the case that nature changes people. (Asahi Shimbun 2013: n.p.)

The gestalt shift enabled by 3/11 meshed with Sakaguchi's previous project, which focused on housing to interrogate the relationship between work, home ownership, and social relationships in contemporary Japanese society. Drawing inspiration from the self-reliance displayed by homeless men in Tokyo, Sakaguchi constructed a number of mobile houses with low cost materials and exploited legal loopholes, such as unresolved land claim disputes that effectively render pockets of land ownerless, to extract himself from the traditional state–citizen relationship. Arguing that this move enabled people to live within a different layer in the city, Sakaguchi used his "0yen" houses to look at how different approaches to housing produces alternative frameworks for conceptualizing public space and ultimately to alternative social relationships and value systems. In Lefebvre's terms, the 0yen house project interrogated the ways in which lived space is produced through the complex interaction of perceived (material, concrete) space, and conceived (mentally constructed) space (Purcell 2002), and his independent state can be seen as a radicalization of this initial investigation.

After the events of March 2011 Sakaguchi's activities scaled up to include bordering processes. In response to the perceived failure of the state apparatus to provide for people in the disaster zone, Sakaguchi declared a state of emergency and relocated to the city of Kumamoto in Japan's southernmost island, Kyūshū. Here he declared himself the head of an independent state (*dokuritsu kokka*) and offered retreats for children displaced by the Fukushima meltdown. In order to legitimize this new polity he draws upon international legal conventions, such as the Montevideo convention on the recognition of states, which, in Article 1, declares that: "The state as a person of international law should possess the following qualifications: a) a permanent population; b) a defined territory; c) government; and d) capacity to enter relations with other states" (http://www.cfr. org/sovereignty/montevideo-convention-rights-duties-states/p15897). Sakaguchi fulfills the conditions of the Montevideo convention, by designating his 31,248 (at time of writing) Twitter followers as the permanent population (a), setting up a seat of government from an old house in Kumamoto (c), and travelling abroad to give lectures on the 0yen and mobile house project (d). He has also released his contact details, including bank account details, via his website and Twitter.

To help facilitate shifts in conceptual space, Sakaguchi has set up a website called "zeroPublic," and it is here that an important part of his borderwork takes place. Here citizens can mark plots of land in Japan that, due to legally ambiguous ownership status or simply by virtue of being empty, may be included in the new state territory (http://www.zero-public.com/). This has the effect of diffusing Sakaguchi's border throughout the physical space of Japan and the virtual space of the Internet, which as a result offers a conceptual challenge to the state's monopoly over the production of space. It also has the effect of connecting non-proximate parcels of land distributed across Japan by virtue of the process of bordering, a process radicalized by the fact that all of Sakaguchi's citizens are authorized to speak in the name of his state and produce territory via bordering. Indeed, the fact that the production of borders has been delegated to people spread across Japan is a further indication of the potential for connectivity inherent in bordering practices. This being said, Sakaguchi recognizes that expanding his territory is, at least potentially, a challenge to the established Japanese state. However, again drawing on his concepts of layers, he argues that by using the land but rejecting the right to own it, his citizens can "escape from the current government's layer" which takes the notion of property rights as a constitutive principle of the polity (http://www.zero-public.com/). By rejecting property rights Sakaguchi's borders become oxymoronic and can simply disappear if the state were to exercise its right to territorial sovereignty. This neatly mirrors, and effectively parodies, state and supra state bordering practices, whereby borders increasingly pop-up whenever they are needed. But the borders also become connective in the sense that they produce open public spaces for chance encounters by virtue of their designation via the website.

In Japan's rapidly ageing, low birth-rate society this attempt to reclaim unused, or as Sakaguchi terms it "left over," property is a significant challenge to the normative life progression of the post-war years. Indeed, by using the border to draw attention to property and land that has fallen through the cracks between institutions involved in the production of lived space in Japan, Sakaguchi challenges naturalized attitudes towards property ownership established in the post-war period (Forrest and Hirayama 2009, Hirayama and Ronald 2008). Therefore this praxis experiment (Perkins and Rumford 2013: 279) in borderwork also highlights the relationship between borders, the production of space, and questions of justice, by demonstrating how they combine to allocate value to people and things within a polity. Returning to Lefebvre's concept of space introduced above, Sakaguchi's playful use of the border can be seen as facilitating a change in lived space by enabling people within the territory of Japan to look again at the physical, perceived space around them unencumbered by the cognitive category of Japan (Brubaker, Loveman, and Stamatov 2004). For example, the goal of owning a new home became firmly embedded within hegemonic life courses in the post-war period (Hirayama and Ronald 2008), and the short term cycle of construction and destruction of housing provides a large stimulus for Japan's construction industry. However, by shedding the cognitive category of Japan, which has within it any number of naturalized prescriptions regarding proper ways of interacting with and placing value on space, the value of empty old property can be reassessed. It is here that concerns regarding the state's response to 3/11 become fused with his previous project.

It is Sakaguchi's invocation of an independent state, and thus a border, that facilitates this interface between different political projects at different scales, making use of international legal frameworks to justify sub-national political action. In this sense Sakaguchi's project is an example of the concept of bordering being used as a political resource (Perkins and Rumford 2013) and to defamiliarize the familiar and imagine different political articulations of space that cut across established notions of state boundedness. But what is the status of the politics in this example? As Amoore and Hall (2010: 311) argue, artistic interventions of this kind raise tricky questions about "how one considers a particular intervention to be political, or to have political effects, without engaging in an exercise of 'what counts' that simply authorizes certain forms of politics." As such it is difficult to measure the impact of his project. He has certainly been successful in terms of generating media interest. His 2012 book was a best seller, and he has gained some traction with the mainstream media, with interviews and review pieces appearing in major newspapers including both the Asahi and Yomiuri. In an opinion piece for the Asahi, philosopher Takahashi Genichiro went as far as likening Sakaguchi's claims to those made by famous figures of Japan's Meiji restoration in 1868, who also called for a new state in a period of acute crisis (Takahashi 2012). Also, Sakaguchi was recently awarded the Second Yoshizaka Architecture Award for work that makes interventions in political and economic life—quite an achievement for an architect who refuses to design buildings (http://www.yosizaka-award.org/).

But to return to the question of politics, it would be easy to dismiss Sakaguchi's project as politically insignificant because it does not fit into "what counts" as politics; similarly it would be easy to dismiss his project as bordering because it is not part of Japan's state bordering practices. However we would be quick to reject this evaluation. Sakaguchi's performative invocation of a state, and with it a border, becomes not only a method of doing politics, but also a way of interrogating what Mouffe (2006) refers to as the "ontological political": the constitutive principles of politics itself. Sakaguchi's borderwork fulfills the important political role of producing alternative imaginaries of public space and political arrangements, which in turn provoke a reconsideration of the naturalized processes that sit in the background of our daily lives.

Sakaguchi is also involved in imagining a new scalar politics. Sakaguchi's state that rejects property rights, justifies itself via international conventions, and devolves bordering practices to its citizens (all of which go on within an already established nation-state) subverts the nested hierarchies of the Westphalian system by making horizontal connections and the spontaneous production of space possible. Therefore Sakaguchi's ad hoc state draws attention to the contingency of sovereignty, state territory, and established bordering practices. To quote Amoore and Hall (2010: 314) once more, this interruption of the everyday is "the true potential of art's political provocation; to make us notice and look again so that nothing is quite as it seems."

Concluding Remarks

This chapter has sought to bring attention to, and importantly push forward an agenda for, a border studies that embraces vernacularization. Such an agenda shifts emphasis away from the state as the primary borderer of choice and draws attention to other, equally important, borders and actors doing their borderwork well away from territorial edges and peripheries. Likewise, seeing from the border, rather than looking across borders, or seeing like a state, allows researchers to take into account the integral relationship between less visible and less traditional borders, borderers as well as those who are bordered. Such an agenda also brings connectivity to the forefront of study in such a way that departs from observations of contact and cooperation within what Martinez (1994) would term "interdependent" or "integrated" borderlands defined and framed as they are by recognizable (and benign) state borders. Taking into account non-state borderworkers doing their borderwork in non-traditional locations, different types of borders—state and non-state alike—can be conceptualized as engines or tissues of non-proximate connectivity, placing borders more centrally within wider processes of global connectivity. In the final section of the chapter, our three dimensions of vernacularization cumulated with a focus on Sakaguchi's borderwork in Japan. The example highlighted how borderwork could be employed as a political resource in order to challenge institutional failings and inadequacies by the Japanese government in light of the March 2011 triple disaster. By evocatively promoting a diffuse and horizontal (non-)state—connecting non-proximate parcels of land within and across the conventional state of Japan—and rejecting the institutionalized bordering between public and private property ownership often attributed to the traditional Westphalian state model, we showed how Sakaguchi's borderwork challenges and subverts the ability of the state to fix conventional space and associated scale hierarchies. In other words, focusing on border vernacularization (and seeing from the border), has shown how Sakaguchi's protest has used borderwork to produce alternative, more spontaneous imaginaries of space that uncover and problematize seemingly naturalized and institutionalized bordering processes that permeate our daily lives. Overall, and in this regard, we hope that putting forward vernacularization as an agenda or framework for border studies creates a substantial foundation for researchers to engage in their own borderwork across the discipline.

References

Amoore, L. 2006. Biometric Borders: Governing Mobilities in the War on Terror. *Political Geography*, 25(3), 336–51.

Amoore, L. and de Goede, M. 2008. Transactions after 9/11: The Banal face of the Preemptive Strike. *Transactions of the Institute of British Geographers*, 33(2), 173–85.

Amoore, L. and Hall, A. 2010. Border Theatre: On the Arts of Security and Resistance. *Cultural Geographies*, 17(3), 299–319.

Amoore, L. and Hall, A. 2013. The Clown at the Gates of the Camp: Sovereignty, Resistance and the Figure of the Fool. *Security Dialogue*, 44(2), 93–110.

Asahi Shimbun. 2013. "Hen" 2013 Sakaguchi Kyōhei-san, Kijima Keiko-san, Unno Motoo-san = Okotowariari. 10th June.

Balibar, E. 2002. *Politics and the Other Scene*. London: Verso.

Balibar, E. 2004. *We, the People of Europe? Reflections on Transnational Citizenship*. Princeton: Princeton University Press.

Bauder, H. 2011. Toward a Critical Geography of the Border: Engaging the Dialectic of Practice and Meaning. *Annals of the Association of American Geographers*, 101(5), 1126–39.

BBC 2012. Japan Quake: Loss and Recovery in Numbers [online]. Available at: http://www.bbc.co.uk/news/world-asia-17219008.

Boer, I.E. 2006. *Uncertain Territories: Boundaries in Cultural Analysis*. Amsterdam, NY: Rodopi.

Brenner, N. 2004. New State Spaces: Urban Governance and the Rescaling of Statehood. Oxford: Oxford University Press.

Brubaker, R., Loveman, M., and Stamatov, P. 2004. Ethnicity as Cognition. *Theory and Society*, 33(1), 31–64.

Bude, H. and Durrschmidt, J. 2010. What's Wrong with Globalization? Contra Flow Speak—Towards an Existential Turn in the Theory of Globalization. *European Journal of Social Theory*, 13(4), 482–500.

Cabinet Office 2007. Security in a Global Hub: Establishing the UK's New Border Arrangements [online]. Available at: http://webarchive.nationalarchives.gov. uk/20071204130052/http://cabinetoffice.gov.uk/upload/assets/www.cabinet office.gov.uk/publications/reports/border_review.pdf.

Casas-Cortes, M. and Corbarrubias, S. 2010. Drawing Escape Tunnels through Borders. *An Atlas of Radical Cartography*, edited by L. Mogel and A. Bhagat. Los Angeles: Journal of Aesthetics and Protest Press.

Cooper, A. (forthcoming). *Rethinking Borders: Globalization, Bordering, Connectivity*. London: Routledge.

Cooper, A. and Rumford, C. 2011. Cosmopolitan Borders: Bordering as Connectivity. *The Ashgate Companion to Cosmopolitanism*, edited by M. Rovisco and M. Nowicka. Aldershot: Ashgate, 236–255.

Cooper, A. and Perkins, C. 2012. Borders and Status-functions: An Institutional Approach to the Study of Borders. *European Journal of Social Theory*, 15(1), 55–71.

Cooper, A. and Rumford, C. 2013. Monumentalising the Border: Bordering through Connectivity. *Mobilities*, 8(1), 107–24.

Durrschmidt, J. and Taylor, G. 2007. *Globalization, Modernity and Social Change*. Houndmills: Palgrave.

Forrest, R. and Hirayama, Y. 2009. The Uneven Impact of Neoliberalism on Housing Opportunities. *International Journal of Urban and Regional Research*, 33(4), 998–1013.

Gschrey, R. 2011. Borderlines: Surveillance, Identification and Artistic Explorations along European Borders. *Surveillance & Society*, 9(1/2), 185–202.

Häkli, J. and Kaplan, D. 2002. Learning from Europe? Borderlands in Social and Geographical Context. *Boundaries and Place: European Borderlands in Geographical Context D*, edited by D. Kaplan and J. Hakli. Lanham: Rowman and Littlefield.

Hedetoft, U. 2003. *The Global Turn: National Encounters with the World*. Aalborg: Aalborg University Press.

Hirayama, Y. and Ronald, R. 2008. Baby-boomers, Baby-busters and the Lost Generation: Generational Fractures in Japan's Homeowner Society. *Urban Policy and Research*, 26(3), 325–342.

Kolossov, V. 2012. Euroborderscapes: State of the Debate. Report 1, European Commission.

Konrad, V. and Nicol, H. 2008. *Beyond Walls: Re-Inventing the Canada–United States Borderlands*. Aldershot: Ashgate.

Kopper, A. 2012. The Imaginary of Borders: From a Coloring Book to Cezanne's Paintings. *International Political Sociology*, 6, 277–293.

Kramsch, O.T. 2011. Along the Borgesian Frontier: Excavating the Neighbourhood of "Wider Europe." *Geopolitics*, (16)1, 193–210.

Lahav, G. and Guiraudon, V. 2000. Comparative Perspectives on Border Control: Away from the Border and Outside the State. *The Wall Around the West: State Borders and Immigration Controls in North America and Europe*, edited by P. Andreas and T. Snyder. New York: Rowman and Littlefield.

Leitner, H. 2004. The Politics of Scale and Networks of Spatial Connectivity: Transnational Interurban Networks and the Rescaling of Political Governance in Europe. *Scale and Geographic Inquiry*, edited by E. Sheppard and R.B. McMaster. Oxford: Blackwell.

Lyon, D. 2012. Introduction. *Liquid Surveillance: A Conversation*, edited by Z. Bauman and D. Lyon. Cambridge: Polity Press.

Luscombe, R. 2009. "Patrol watches Texas–Mexico border from pub in Australia." *The Guardian*, 23rd March. Available at: http://www.guardian.co.uk/world/2009/mar/23/texas-mexico-patrol-webcam-australia.

Martinez, O.J. 1994. The Dynamics of Border Interaction: New Approaches to Border Analysis. *Global Boundaries: World Boundaries*, edited by C.H. Schofield. London: Routledge, 1–14.

Massumi, B. 1993. *The Politics of Everyday Fear*. Minneapolis: University of Minnesota Press.

Mignolo, W.D. 2000. The Many Faces of Cosmo-polis: Border Thinking and Critical Cosmopolitanism. *Public Culture*, 12(3), 721–48.

Multiplicity 2005. Borders: The Other Side of Globalisation. *Empires, Ruins + Networks: The Transcultural Agenda in Art*, edited by McQuire and N. Papastergiadis. London: Rivers Oram Press.

Newman, D. and Paasi, A. 1998. Fences and Neighbours in the Postmodern World: Boundary Narratives in Political Geography. *Progress in Human Geography*, 22(9), 186–207.

Papadopoulos, D., Stephenson, N., and Tsianos, V. 2008. *Escape Routes: Control and Subversion in the 21st Century*. London: Pluto Press.

Perkins, C. and Rumford, C. 2013. The Politics Of (Un)Fixity and the Vernacularization of Borders. *Global Society*, 27(3), 267–282.

Purcell, M. 2002. Excavating Lefebvre: The Right to the City and its Urban Politics of the Inhabitant. *GeoJournal*, 58(2–3), 99–108.

Rosière, S. and Jones, R. 2012. Teichopolitics: Re-considering Globalisation Through the Role of Walls and Fences. *Geopolitics*, 17(1), 217–234.

Rovisco, M. 2010. Reframing Europe and the Global: Conceptualizing the Border in Cultural Encounters. *Environment and Planning D: Society and Space*, 28(6), 1015–30.

Rumford, C. 2006. Borders and Rebordering. *Europe and Asia Beyond East and West: Towards a New Cosmopolitanism*, edited by G. Delanty. London: Routledge.

Rumford, C. 2007. Does Europe have Cosmopolitan Borders? *Globalizations*, (4)3, 327–339.

Rumford, C. 2008. Introduction: Citizens and Borderwork in Europe. *Space and Polity*, (12)1, 1–12.

Rumford, C. 2011. Seeing like a Border (contribution to symposium "Interventions on Rethinking 'The Border' in Border Studies"). *Political Geography*, 30(2), 61–69.

Rumford, C. 2012. Towards a Multiperspectival Study of Borders. *Geopolitics*, 17(4), 887–902.

Rumford, C. (forthcoming). Towards a Vernacularized Border Studies: The Case of Citizen Borderwork. *Journal of Borderlands Studies*.

Sakaguchi, K. 2012. *Dokuritsu Kokka no Tsukurikata*. Tokyo: Kōdansha.

Scott, J. 1998. *Seeing like a State: How Certain Schemes to Improve the Human Condition Have Failed*. New Haven, CT: Yale University Press.

Shun'ya, Y. and Loh, S.L. 2012. Radioactive Rain and the American Umbrella. *The Journal of Asian Studies*, 71(02), 319–331.

Strüver, A. 2004. Everyone Creates One's Own Borders: The Dutch–German Borderland as Representation. *Geopolitics*, 9(3), 627–48.

Takahashi, G. 2012. Shinshikō wo Hagukumu, Kuni ya Kenpō, Jibun de Tsukkuchae. *Asahi Shimbun*, 26th July.

van Houtum, H., Kramsch, O., and Zierhofer, W. 2005. *B/ordering Space*. Aldershot: Ashgate.

van Schendel, W. 2005. How Borderlands, Illicit Flows, and Territorial States Interlock, *Illicit Flows and Criminal Things: States, Borders, and the Other Side of Globalization*, edited by W. van Schendel and I. Abraham. Bloomington: Indiana University Press, 38–67.

Walters, W. 2006. Border/Control. *European Journal of Social Theory*, 9(2), 187–204.

Walters, W. 2009. Anti-political Economy: Cartographies of "Illegal Immigration" and the Displacement of the Economy. *Cultural Political Economy*, edited by J. Best and M. Paterson. London: Routledge, 113–138.

Wastl-Walter, D. 2011. *The Ashgate Research Companion to Border Studies*. Farnham: Ashgate Press.

Chapter 3

Policing Borders, Policing Bodies: The Territorial and Biopolitical Roots of US Immigration Control

Mathew Coleman and Angela Stuesse

Introduction: Lessons from IRCA

In 1981, the bipartisan US Select Commission on Immigration and Refugee Policy (SCIRP)—also known as the Hesburgh Commission—published their much anticipated *US Immigration Policy and the National Interest*. The report detailed the impact of refugee policy, legal immigration quotas, and US-bound undocumented immigration on what lawmakers described as "social, economic, and political conditions in the United States," "demographic trends," "the conduct of foreign policy," and "present and projected unemployment in the United States." The report makes for strange reading on account of the many interests that were given voice in the document. Martin (1982), for example, described the report shortly after its release as a largely uncommitted "middle position" on what had been said to date by both lawmakers and researchers on the topic of immigration and refugee policy—a problem he attributed to the commissioners' broad disagreement over key issues such as the demographic and economic consequences of US immigration policy. And yet despite its rehashed and sometimes contradictory nature, the report did maintain a clear focus on global population growth and on what global population pressure could mean for the future of the US as a traditionally white, Anglo-Saxon settler colony. For example, the report dwelled at length on "the fact that we live in a shrinking and interdependent world" and that "tidal movements of people ... propelled by economic forces" necessitate "closing the back door" to undocumented immigration (US Select Commission on Immigration and Refugee Policy 1981b: 3, 11). In order to make the case for "closing the back door," the report enumerated "serious adverse effects" posed by a growing "fugitive underground class" of undocumented immigrants driven to the US by population-related economic dysfunctionality in Mexico and Central America. These included ethnic tension, cultural enclavism, a possible loss of civic culture and cohesion, job displacement, wage depression, and rising healthcare as well as education costs (US Select Commission on Immigration and Refugee Policy 1981b: 11). Most alarmingly, the report argued that widespread lawlessness was a likely effect of uncontrolled undocumented migration:

> The presence of substantial numbers of undocumented/illegal aliens in the United
> States has resulted not only in a disregard for immigration law but in the breaking
> of minimum wage and occupational safety laws, and statutes against smuggling
> as well. As long as undocumented migration flouts U.S. immigration law, its
> most devastating impact may be the disregard it breeds for other U.S. laws. (US
> Select Commission on Immigration and Refugee Policy 1981b: 42)

As the commission wrote elsewhere, it was only via a comprehensive legalization program that so-called "shadow" populations of undocumented immigrants and the broader rule of law could be rescued (see "Out of the Shadows: The Rule of Law Applied" in US Select Commission on Immigration and Refugee Policy 1981a: 631–664).

The Hesburgh Commission's strong language on undocumented immigration played a very important role in setting immigration legislation during the 1980s (Zolberg 2006: 337–381). In particular, the Commission's endorsement of a three-pronged approach to fending off the impending legal and demographic consequences of US-bound undocumented immigration—enhanced border policing, workplace enforcement, and selective legalization—prompted Senator Alan Simpson (R-WY) and Representative Romano Mazzoli (D-KY) to re-open the immigration reform effort spearheaded by Representative Peter W. Rodino (D-NJ) during the Ford and Carter administrations. The Commission's triplet of recommendations became the backbone of Simpson and Mazzoli's Immigration Reform and Control Act (IRCA), signed into law by President Reagan in November 1986.

IRCA was notable on several grounds. It was the first major immigration law reform bill to move successfully through both the House and the Senate in the wake of the 1952 McCarran-Walter Act, and thus stands as one of the legislative cornerstones of the post-Chinese Exclusion Act period of immigration lawmaking in the US. IRCA was also the first piece of legislation in the US to deal in a substantive fashion with undocumented immigration. But perhaps most remarkable about the law was its nearly singular focus on population as a problem deserving of sustained state intervention. As Chock (1995) argues, the entire IRCA edifice hinged on the problem of population. Population was a catch-all category used broadly by lawmakers during the lead up to IRCA to outline the apparently biological—or asocial and ahistorical—nature of undocumented immigrants' political, economic, and social practices, as well as the problem of cultural discordance between immigrant and citizen groups and competition between them for scarce resources. Although key lawmakers like Senator Simpson may have championed the merits of race-free, class-free, and gender-free debate about undocumented immigration, Chock shows that their engagement with population in natural history terms, particularly when discussing the problem of immigrant acculturation and lawfulness, provided a proxy vocabulary for an oftentimes explicitly raced, classed, and gendered account of the various social ills and disturbances associated with undocumented immigration. Indeed, as Coleman

(2008) has outlined elsewhere, IRCA was authored on the back of a wide-ranging, mostly conjectural, biopolitical inventory of resident undocumented immigrants' use of scarce schooling, healthcare, and entitlements resources; resident undocumented migrant populations as potential disease vectors; and, resident undocumented migrants' ethnic and linguistic separatism as a threat to the so-called "melting pot" approach to citizenship, as well as to core public values and practices associated with "American" culture.

It is now commonplace for both supporters and detractors of amnesty to recall IRCA as an overly generous "clean slate" immigration bill. From this standpoint, IRCA was an undiscriminating, quick fix legalization law whose major policing focus was not on amnesty per se but on increasing the Border Patrol presence along the US–Mexico border so as to prevent future growth in the resident undocumented population. This is true to the extent that IRCA kicked off a period of substantial growth in immigration enforcement budgets, most of which was directed at Border Patrol resources along the US–Mexico border (Dixon and Gelatt 2005, Dávila, Pagán, and Grau 1999). Indeed, IRCA can be understood as an important first major salvo in the militarization of the US–Mexico border under the explicit banner of immigration control. For example, in the spirit of the Hesburgh Commission's emphasis on expanding Border Patrol resources as a necessary condition of amnesty, IRCA legislated nearly a billion dollars of additional funding for the Immigration and Naturalization Service (INS) for fiscal years 1986 and 1987 (see budget data in US House of Representatives Judiciary Committee 1986). The bulk of this budget, some 57 percent, was dedicated to border enforcement. This money funded a major expansion in the ranks of the Border Patrol, new helicopters for most Border Patrol sectors, new surveillance equipment at the US–Mexico border, and additional Border Patrol stations, detention facilities, as well as road checkpoint installations across the US Southwest (Meyers 2005).

This said, IRCA comprised much more than border enforcement. As Gonzalez-Baker (1997) argues, IRCA struck a balance between the "clean slate" border-centric approach and a "firm equities" approach to legalization, meaning that, in addition to beefing up the border, the law engineered a carefully policed pathway to regularization, and eventually citizenship, for resident undocumented immigrants. In other words, IRCA's amnesty process was highly selective and corrective, rather than a simple blanket reprieve. For example, the law did not apply to all undocumented immigrants but only those deemed lawful, healthy, and economically productive. IRCA barred legalization as well as employment authorization for immigrants convicted of a crime of moral turpitude, a drug offense, any felony with a term of imprisonment exceeding one year, and/or three misdemeanors at any point in their lifetime. It also excluded a range of immigrants on public health grounds, including HIV/AIDS, tuberculosis and syphilis. And perhaps most importantly, the law was conspicuously shaped by what has been called "balanced budget conservatism" (Calavita 1996). Significant congressional attention was given to ensuring that migrants legalized under IRCA would be economically productive and not a drain on taxpayers. This took two principal forms. On the one hand, migrants likely to become a public charge—as measured by food stamp and other aid use—were for the most part denied legalization. Lawmakers also ensured that "continuous work documentation preference," as it was termed, would be a threshold for legalization, which disadvantaged mostly female undocumented workers employed in the underground domestic services industry (Arp, Dantico, and Zatz 1990). Moreover, IRCA made it more or less impossible for migrants, once granted work authorization, to become a ward of the state, as it prohibited—with some exceptions for the elderly, pregnant, and disabled, and in the event of a life-threatening emergency—newly legalized migrants from drawing on aid and services for a five-year temporary residence trial period. In sum, what "firm equities" meant was that even if the initial steps in the legalization process were based on a simple statutory requirement and were thus not per se discretionary, later steps included important regulative caveats and bars-to-admission that first effectively sorted the population of undocumented immigrants into those eligible and ineligible for work permits, and second, closely regulated this population on criminal, health, and economic grounds once admitted.

We start this chapter off with the above brief recap of the Hesburgh report and IRCA on account of how we understand the immigration reform effort during the early 1980s as focused on policing bodies rather than simply borders. For us, IRCA highlights in a particularly clear manner the biopolitical aspects of US immigration control. By biopolitics we mean techniques of government which seek to minimize certain forms of risk or uncertainty at the level of dynamics and exchanges within a population, and not simply territorially. This is not how IRCA, and indeed US immigration control policy throughout the 1990s, is usually interpreted. The dominant critical reading of post-IRCA US immigration policy is that lawmakers, concerned about US borders as sites of uncontrolled undocumented entry, funded their militarization. Political geographer Joseph Nevins makes this point in his groundbreaking account of the rise of the "gatekeeper state" at the US–Mexico border during the 1990s (Nevins 2002). As Nevins puts it, a general consensus regarding the "geographical crime" of undocumented entry—literally, migrants' disregard for the territorialization of the law—enabled lawmakers, during the late 1980s and 1990s, to gradually militarize the US–Mexico border as well as criminalize undocumented immigration (Nevins 2002: 141). From this general theoretical position (see also Dunn 1996, Andreas 2000), IRCA and the decade-long border build-up that the law helped spur is to be understood primarily in geopolitical terms as a military-territorial partitioning and/or policing of US borders. Geopolitics, as in controlling access to strategically occupied compartments of topographical space, is clearly part of IRCA's genetic make-up, as well as part of its legacy. For example, as noted above, IRCA reinvigorated a series of externally-oriented interdiction activities at US borders, and specifically at the US–Mexico border. Moreover, that US borders were overwhelmed by unpoliced, undocumented entry was certainly a very common refrain for lawmakers throughout the frenzied period of immigration lawmaking, bookended on the one hand by the formation of SCIRP and on the other hand by IRCA in 1986. And still more generally, we agree firmly that the perceived sanctity of legal-territorial borders in an abstract rule-of-law sense is an extremely important touchstone for immigration reform lawmakers and their publics (Purcell and Nevins 2005). At the same time, our brief summary of IRCA as concerned with the demographic and legal impact of "out of control" undocumented immigrant populations is meant to point out that less explicitly territorial concerns can be woven into legal-territorial justifications for immigration law reform. Indeed, what strikes us as most important about IRCA is how the law, but certainly US immigration policy more generally, focused on immigration control as doubly a problem of unpoliced borders and unpoliced populations. This leads us to the position that geopolitics and biopolitics, particularly in the context of research on US immigration enforcement, need to be understood as deeply intertwined, as opposed to distinct, technologies of governance.

In this chapter we seek to close the gap between what we see as competing geopolitical and biopolitical analyses of US immigration control policy. We see this debate as reflecting a more general impasse across the social sciences and humanities which insists on biopolitics and geopolitics as opposed models of power. Although both can be understood as mechanisms employed by states to cope with threat and (in)security, geopolitics and biopolitics are usually differentiated on account of their respective concerns for space and time, or what we describe here as topography and topology. For example, whereas geopolitics is typically characterized as a strategy of territorial control focused on the fortification of state-territorial borders (see critique in Agnew 1994), biopolitics is usually defined as an abandonment of spatial or territorial control per se due to its focus on the "everywhere" management of populations. Indeed, in place of the geopolitical focus on partitioning and policing territorial space, biopolitics is usually conceptualized as a way of dealing with an anticipated but never quite known calendar of possibly destabilizing and/or threatening events, anywhere. In other words, whereas geopolitics has been conceptualized as a locational approach to security—in terms of a defensive array of barriers to movement, membership, exchange, circulation, etc.—biopolitics has been conceptualized non-locationally as ensuring that over time certain practices—deemed undesirable—are made less and less viable. We hope to undermine this distinction, and in so doing provide a hybrid geopolitical and biopolitical account of US immigration control today.

Our chapter comprises two basic sections. In a first section, we review two approaches to the study of borders, which we label "topographical" and "topological." Our goal in this section is to point out that while each approach has its merits they are largely missing out on what the other has to contribute to border studies. In a second section, we focus our argument on present-day immigration control policy in the US. We argue that the growth of interior enforcement, away from US territorial borders such that any number of routine practices and/or encounters can result in detention and deportation for undocumented immigrants, can be described as a form of biopolitics in the sense that its strategic goal is to destabilize the lived everydays of immigrant laborers to the point that social reproduction is impossible. But this does not mean that geopolitical borders are unimportant to immigration enforcement. Indeed, our argument is that the topographical/geopolitical tactic of making the US–Mexico border increasingly hard to cross is an ongoing precondition for the topological/biopolitical strategy of radically increasing levels of day-to-day uncertainty for immigrant populations.

Theorizing Borders: Topography, Topology, and Biopolitics

Border studies is not a coherent enterprise. One reason why this is the case is because the field is broadly interdisciplinary. As anthropologists Donnan and Wilson suggest in a recent survey, the fact that border studies scholars come from across the humanities and social sciences, and that they bring with them a large variety of disciplinary best practices, has earned border studies a "fashionable branding" in academia. But, as they go on to argue, border studies' interdisciplinarity has also tended towards "an uncritical accumulation or juxtaposition of different perspectives which in itself did not advance the study of borders very far" (Wilson and Donnan 2012: 15). We would add to this problem that border studies has, until quite recently, been a mostly empirical undertaking that has prioritized a catalogue-style approach to border research at the expense of more theoretically-minded claims about how borders work, and why. There are some notable exceptions to this claim. One very important case in point is Gloria Anzaldúa's groundbreaking account of borders as infrastructures built from contradictory moments of resistance and reconstruction—and moreover her insistence that borderlands exceed state territorial borders as such (Anzaldúa 1987). We note also Jean Gottmann's pioneering work on borders as psychosomatic geographies caught between what he described as opposing geopolitical and geoeconomic logics of closure and openness, respectively (Gottmann 1973). More generally, we recognize the significant work of Mexican American/Chicano Studies scholars theorizing the US–Mexico cultural borderlands and the power dynamics that saturate everyday life there, too often overlooked by those ensconced in disciplinary silos (see, for example, Flores 2002, Limón 1994, Paredes 1958, Pérez 1999, Rosas 2012, Saldívar 1997, Trujillo 2009, Vélez-Ibañez 1996). But for the most part border studies has shied away from this sort of theoretical work, and has instead adopted an area studies approach (see discussion in Paasi 2005, Newman 2006, Newman and Paasi 1998, Kolossov 2005).

Border studies' general neglect of theory is, of course, now very much in question. Although the descriptive approach to borders remains an important part of the field, border studies began to engage in much more theoretically-minded research in the late 1990s (Newman 2002). A major theme during this period was on the importance of borders as "socially constructed" by groups as well as crucial to individual and group identity formation (e.g. Paasi 1996). As a result, it is now much more commonplace for border studies scholars to draw on a range of theories regarding power, political economy, and social difference to generalize about borders. It is also commonplace for border scholars to be at the forefront of exciting new theoretical innovations on these topics, and moreover for border theory to be at the center of broader debates about empire, law, neoliberalism, and so on. What interests us specifically about the recent theoretical turn in border studies is the disagreement that it has spawned about borders, which as objects of academic scrutiny are arguably more in question today than they were during the reign of descriptive research. Indeed, in the wake of its largely empiricist as well as idiographic past, borders themselves have emerged as "essentially contested concepts" (Connolly 1974) in border studies scholarship.

Our point in raising the problem of borders as "essentially contested" is not to clear ground in the name of a new general theory of borders, or to take sides. Rather, we see the complexity and openness of the current debate around borders (Jones 2009, Rumford 2006) as an opportunity to engage in creative work across competing schools of thought concerning specifically how borders work, and crucially *where* we might go about finding them (see various viewpoints on offer in Johnson et al. 2011). Perhaps too crudely we see two primary theoretical approaches to borders in border studies today. The first approach is topographical. By this we mean that borders are theorized as territorially concrete and identifiable measures of space. From this standpoint, borders are geopolitical-territorial lines of division that describe units of space (or territories) in relation to one another, which are themselves characterized by some form of measurable spatial extent or reach. This is not to say that a topographical account of borders treats the latter, or the spaces that they divide, as quasi-natural realities (Fall 2010). Rather, to approach borders topographically is to insist that spaces of differentiation and segmentation, if social as well as political economic, are nonetheless materially locational in some robust sense. Indeed, a topographical approach to borders would conceptualize them as relatively durable architectures without succumbing to the "territorial trap" of seeing them as the fixed outer edges of some apparently natural social, political, and/or economic container (Agnew 2008).

This important distinction between topography and spatial fixity reflects a larger concern for the production of place inherent in the concept of topography. For example, Katz explains topography as a way of conceptualizing the local and non-local aspects of place-making as well as the ongoing conflicts integral to this process—which, as she sees it, allows for a dynamic theorization of place amidst a larger uneven geography of flows and connections (Katz 2001: 1214, for a similar appreciation of topography, see Staeheli and Nagel 2006, Nagar et al. 2002, Martin 2005). In this sense, topography does not embrace the idea of place as an essential location, in contrast to the abstract place-spanning force field called space (see Massey 1991). Nonetheless, in the border studies context, we would argue that the topographical approach tends to conceptualize borders as some of the most located, most concretized, and most durable of social geographies. Even as this literature in general problematizes borders as "social constructions" of "insideness" and "outsideness" rather than as natural entities and some have insisted we recognize the hybridity and heteroglossia of borderlands, borders are on the whole understood from the topographical standpoint in terms of a relatively straightforward territorial ontology of inside/outside—and as such are remarkably muscular "social constructions." Moreover, the significant majority of the topographical border studies research has, in our opinion, proceeded much too narrowly in terms of what counts as a border. The focus in this work has been overwhelmingly on one particular category of border and its affirmation of the inside/outside binary: the geopolitical, and sometimes also geophysical, borders separating states.

We understand the topographical tendency to see borders as constructed and yet enduring points of division (and connection) between specifically state territories as, in large measure, a knock-on effect of the critical geopolitics and critical international relations revolutions, which started during the mid-1980s (see overview in Kuus 2010). Critical geopolitics and critical international relations, despite some considerable differences of opinion, collectively urged a rethinking of, first, the state as a uniform and bounded space, and second, state power as a rational enterprise. Although the emphasis in this debate was on upsetting established models of state power in mainstream political geography and international relations theory, the problem of state borders was always at hand. For example, critical geopolitics and critical international relations scholars pushed for borders to be recast not as simple pedagogic givens in the international space "between" states but, crucially, as part of the performative infrastructure of states (Newman 2003). Or, as Agnew noted in the aftermath of the critical geopolitics revolution, borders "are not simply practical phenomena that can be taken as a given. They are complex human creations that are perpetually open to question" (Agnew 2008: 2). And yet, if we look at political geography and political science literature on borders authored in response to these post-Cold War theoretical innovations, we see very familiar objects of analysis rather than geographies "open to question." A good example of this is the theoretically-minded research identified with the "Territory, Identities, Movement" as well as "Identities, Borders, Orders" research groups (Brock 1999, Albert and Brock 1996, Agnew 1999, Albert, Jacobson, and Lapid 2001). This scholarship very usefully undermined conventional approaches to the state and its borders by insisting, among other things, that borders are socially created and maintained. At the same time, this body of research re-centered our attention on precisely inter-state landscapes as preeminently important locations for thinking through the precarious materialization of "world-constituting distinctions ... such as inside/outside, anarchy/hierarchy, domestic/foreign, self/other, here/there, and so on" (Lapid 2001: 11). Put slightly differently, although this literature allowed us to shake loose both an essentialist and primordialist approach to borders by querying the unthought inside/outside ontology at the heart of disciplines like political geography and international relations, its focus on criticizing this "territorialist" ontology paradoxically led to a repetition of borders in terms of the problem of state spatiality, albeit under the decidedly less glorious rubric of "social construction." Moreover, we note more generally that some of the most important work today in border studies theorizes borders primarily in terms of the external bounding processes constitutive of state spatiality (Diener and Hagen 2010, Donnan and Wilson 2010, Brunet-Jailly 2007). This is, to be clear, not to underplay the importance of borders as locationally robust infrastructures which claim, rather than simply represent, an inside/outside threshold. For example, in the US–Mexico case, which we know better than other cases, the border between these two countries is without doubt a very significant marker of social difference, belonging, and political economic privilege. But it is not the only border at work in terms of the US–Mexico relationship. Moreover, we would argue that the US–Mexico border is matched in its intensity and violence by other sorts of more mobile borders "on the inside" of the US. We will look in more detail at the US–Mexico case study in the following section.

A second, much less dominant, approach to the theoretical study of borders foregrounds the problem of topology. Topology, broadly, refers to non-planar, non-linear, non-territorial, and non-distance based accounts of space and place, and their production (Belcher et al. 2008). A topological approach to the problems space and place—the former usually interpreted as general and intrusive, and the latter as grounded and bounded—de-differentiates the two such that the inside/outside ontology at the heart of the space/place distinction dissolves. As Amin explains it, topology concerns "geographies constituted through ... folds, undulations, and overlaps that natural and social practices normally assume, without any a priori assumption of relations nested in territorial or geometric space" (Amin 2002: 389, see also Allen 2009). The concept of topology boils down to three basic claims. First, the usual priority placed on propinquity, or nearness, when it comes to understanding what happens in a certain site is replaced by an emphasis on non-metric connectivity (Amin 2004). Second, topology draws attention to the transductive or in-formation quality of social relations in specific sites (Dodge and Kitchin 2005). Indeed, topology has been at the center of recent debates in political geography on the relationship between power and space, largely on account of the ascendency of Actor Network Theory's insistence on the contingencies of socio-spatial practice in particular locations (Latour 2005, Thrift 2008). Third, and most important for us, topology throws borders into question as precisely located infrastructures. The basic point of the topological approach, especially for those interested in the coupled problems of power and bordering, is that there is indeed no necessary coupling. The concept of topology suggests that there is no static and stable material domain called "the social" which can be meaningfully divided into units or chunks of space, either vertically in terms of "scale" or horizontally in terms of "regions," and which can then be used to theorize either the deployment of power or its effects (Woodward 2010, Marston, Jones, and Woodward 2005).

We note too that there has been an explosion of work on biopolitics in border studies which has tried to destabilize borders in terms of state-based tactics of areal differentiation and the practice of territorial capture. This trend is obviously of a piece with the research on topology, as above, even if the latter is not always cited. For example, borrowing heavily from Michel Foucault's work on the governance of sexuality, scholars have sought to shift the problem of state power away from the surveillance and policing of territory to the mass surveillance and policing of mobile populations. Some of this work uses population alongside territory, in the sense that the governance of populations can be understood as a proxy for territorial control and yet as a different way of calculating space (see, for example, Elden 2007). But by far the bulk of the biopolitics research has posed territory as a disappearing object of governance, and territorial control as an outmoded art of governance. Dillon, to pick one of many possible scholars working in this area, has argued, for example, that Foucault's work on biopolitics "opens up an entirely different spatial configuration of security" based on the problem of circulation rather than interdiction or distribution (Dillon 2007: 11). As Dillon explains, whereas "distribution signals a world to be divided between sovereign territorial political subjects and their competing hegemonies, circulation concerns a world understood in terms of the biological structures and functions of species existence together with the relations that obtain between species life and all of its contingent local and global correlations" (Dillon 2007: 11). Dillon concludes by noting that biopolitics has effected a "shift in the referent object of security from sovereign territoriality to life" (Dillon 2007: 11).

This basic Foucault-inspired distinction between power in terms of territory and power in terms of population, as well as the similarly-spirited move noted above towards topology in border studies, has fueled a range of research on the changing spatial strategies of immigration enforcement, particularly in the US and European contexts. This work points away from borders in terms of states' territorial edges and to the proliferation of "everyday" borders and immigrant policing within states (Walters 2006b, Walters 2006a, Vaughan-Williams 2009, Huysmans 2000, Bigo 2002, Amoore 2005, Parker and Vaughan-Williams 2009). In Bigo's (2001) very provocative turn of phrase, border control has become like a Möbius strip in the sense that it is no longer simply located at states' inside/outside interface, which itself is increasingly difficult to locate. Importantly, the shift from topography to topology, or from territory to population, does not imply a softening of borders; this is not a cousin of the so-called "borderless worlds" research that emerged in the immediate post-Cold War context. If anything, border studies scholars working from the topological and/or biopolitical standpoints have theorized borders as, unlike state-territorial borders, constantly consequential in their prosaic, topological–biopolitical guise. However, it is certainly the case that this research de-emphasizes states' so-called "hard" borders. And here we note a difficult logic of substitution at work in much of the topological–biopolitical work on borders. In our opinion too much of the topological research, in trying to carve out room against a strictly topographical and/or geopolitical-territorial approach to borders, replaces rather than supplements borders qua state-territorial "edges" with the less located problem of topological and/or biopolitical division. We see this logic of substitution more generally in the human geography literature insofar as topographical space is seen more or less calendrically as an antiquated mode of analysis for a now mostly defunct period of socio-spatial reality (see discussion in Coleman 2011, Elden 2011, Latham 2011, Paasi 2011). Rather than overdraw the distinction between the topological and the topographical, and the territorial and the biopolitical, we propose instead investigating their combination. This seems an important move given that, despite the proliferation of new forms of borders and bordering, topographical-territorial borders are far from obsolete (Jones 2012). As Rosière and Jones argue (2012), the "hard" barrier function of borders is arguably now more than ever at work both between and within states, in the form of walls and fences along international borders as much as around gated communities.

Our attempt to tack between a topographical tradition focused intently on problematizing state-territorial borders and a topological tradition focused on elucidating how biopolitical borders work beyond those topographical parameters has been deeply influenced by the new mobilities turn in human geography. In general this literature has insisted on theorizing the problem of enforced immobilization and borders through both topographical and topological lenses, or as a problem of geopolitics and biopolitics (Turner 2007). Rather than rehearse that field and its debates (see Stuesse and Coleman 2014), we want to briefly note here our debt on this question to Sandro Mezzadra and Brett Neilson's work on the precarity of immigrant labor. We are particularly drawn to Mezzadra and Neilson's understanding of immigrant policing as a geographically complex strategy which blends external qua exclusive and internal qua modulating modes of immigration control. The first thing to note is Mezzadra and Neilson's attention to immigrant policing as an obviously geopolitical-topographical problem of state-territorial borders and interdiction practices. For example, Mezzadra and Neilson (2008) emphasize state borders as exclusion-oriented legal-territorial barricades which in a very real sense restrict the movement of laboring bodies between still importantly territorialized labor markets. At the same time, Mezzadra and Neilson note throughout their work that there has been a comprehensive "multiplication of borders," and that as a result immigrant policing cannot be reduced to a geography of walls and fences. Indeed, in addition to what we might think of as their more traditional geopolitical function as spatial barricades, Mezzadra and Neilson allow that borders work biopolitically, at the level of the population, and then topologically, as in unevenly across space and time.

In order to unpack the problem of biopolitical-topological borders, as a supplement to borders in a more straightforward sense, Mezzadra and Neilson draw a parallel between the current detention and deportation regime in Fortress Europe and so-called "benching" in labor market "body shops." The latter refers to a process whereby Indian hi-tech workers abroad are held in reserve by labor market brokers for a period of time so as to raise the price of labor and thereby increase demand (Mezzadra and Neilson 2013: 131–166). Importantly, "benching" is a temporary condition—a "time of forced suspension" or "controlled withdrawal"—which applies to select laborers rather than to an entire population or cohort of workers (Mezzadra and Neilson 2013: 136–137). Indeed, Mezzadra and Neilson characterize benching as a form of "differential inclusion" which selectively excerpts and then reinserts laborers into specific labor markets, in order to manipulate demand. As they note: "Inclusion, in this perspective, is not an unambiguous social good, but a differential system of filtering and stratification that functions as a means of hierarchization and control" (Mezzadra and Neilson 2008: no page numbers). Mezzadra and Neilson's overall point in comparing immigrant policing in Europe with labor market "benching" is twofold. First, they use the example of labor withdrawal at the heart of the "body shop" phenomenon to propose that the practice of detention and deportation does not necessarily contradict the demand for undocumented labor (see also Mezzadra and Neilson 2003 on detention as a "decompression chamber" which places controls on labor market incorporation in the context of widespread capital mobility). This recalls Nevins' earlier analysis of US–Mexico border enforcement as a sort of demographic mop-up in the wake of the US-imposed neoliberalization of Mexican labor markets post-NAFTA (Nevins 2002, see also Calavita 1992), and as such suggests an uneasy intimacy—rather than a stark opposition—between economy and security regarding the policing of borders and immigration more generally. Second, and for our purposes perhaps most important, Mezzadra and Neilson build on the example of "body shops" to propose a model of immigrant labor control deployed *within* the most privileged spaces of global capitalism, or otherwise in places typically (mis)understood as migrant destinations. As they suggest, detention and deportation is a topological system of forced withdrawal from national labor markets which mimics the dis-locating border-crossing geographies of migrants rather than simply the legal-territorial geography of the Westphalian state system (see also Mezzadra and Neilson 2003: 8). Indeed, in the spirit of Balibar's analysis of borders as shifting zones rather than unbroken lines (Balibar 2002: 75–103), Mezzadra and Neilson characterize borders as "cut[ting] across and exceed[ing] existing political spaces." This emphasis on the dis-location of border policing away from states' territorial edges leads to a further suggestion that immigrant policing transcends a simple inside/outside logic of interdiction, or of capture and geographic removal. To make this specific point, Mezzadra and Neilson develop the notion of "temporal borders," which is essentially the "body shop" equivalent of labor market withdrawal but in the realm of detention and deportation (Mezzadra and Neilson 2013: 131–166). Temporal borders are not about uniform and enduring interdiction practices and infrastructures at states' territorial fringes which systematically sort migrating populations according to what has elsewhere been called an "international police of aliens" (Walters 2002). Rather, temporal borders are, like the temporary withdrawal characteristic of "body shops," fitful in their enforcement, or not always turned on. In sum, the comparison with "body shops" allows Mezzadra and Neilson to model immigrant policing as, first, variously located "on the inside," and second, as intermittent. The result is an on-and-off patchwork configuration of immigrant policing aimed at resident immigrant populations which is broadly consistent with the geopolitical policing of borders, even if thoroughly different in its implementation.

Immigration Control Policy in the US, Post-9/11

In August 1994 the Border Patrol announced a new border control strategy in the US Southwest. The approach, dubbed "prevention through deterrence," expanded the use of fencing, vehicle barriers, lighting, and surveillance technology along the US–Mexico border in order to deter would-be undocumented migrants from crossing into the US between official border ports of entry (Nevins 2002, Dunn 1996, Cornelius 1998, Andreas 2000). The "prevention through deterrence" strategy was in large part the product of a 1993 Sandia National Laboratories report commissioned by the White House Office on National Drug Control Policy. The White House had asked the Lockheed Martin subsidiary to complete a "systematic analysis of the security along the United States/Mexico border between the ports of entry and to recommend measures by which control of the border could be improved" (cited in US Government Accountability Office 1995: 11). In their report, Sandia recommended that the Border Patrol build and manage a multiple barrier fence along the most densely urbanized sections of the US–Mexico border in order to discourage would-be border crossers as well as channel border crossers to less built-up sites along the border where the Border Patrol could then engage in mass detention and deportation operations. The Sandia study itself was not entirely novel. For example, in 1990 the San Diego Border Patrol sector, in conjunction with the Department of Defense, built a 14 mile steel fence along the most urbanized section of the California border in order to funnel undocumented migrants away from San Diego. Similarly, in 1993, the El Paso Border Patrol sector refocused its efforts on so-called "linewatch" duty directly at the US–Mexico border. The El Paso strategy was the outcome of a lawsuit against the Border Patrol by students and teachers at Bowie High School, near the Bridge of the Americas linking El Paso and Ciudad Juárez. The lawsuit alleged that Border Patrol officers were using racial profiles to detain suspected undocumented migrants in working class Latino neighborhoods adjacent to the border. As part of the settlement between the Border Patrol and the school, the sector chief ended the practice of neighborhood patrols and focused his resources on deterrent policing directly at the geopolitical border between the US and Mexico (Dunn 2010).

The new "prevention through deterrence" strategy marked a major break with past Border Patrol practice. With the exception of the small-scale experiments in San Diego and El Paso, noted above, the US–Mexico border had not been a priority enforcement site for the Border Patrol. For example, the wall and fence infrastructure that is now all-pervasive at the border was more or less non-existent when the Sandia report was published. Moreover, the Border Patrol's primary enforcement focus had been on apprehending undocumented immigrants in the US interior; preventive enforcement based on the massing of Border Patrol officers and resources between ports of entry on the US–Mexico border was a relatively untested strategy. Indeed, in addition to policing neighborhoods and bus stops near the border for suspected undocumented migrants, the Border Patrol's primary focus had been on inspecting vehicles and passengers during roving traffic enforcement operations away from the border, as well as at fixed and temporary traffic checkpoints on key highways leading north from the US–Mexico border (this continues to be an important aspect of Border Patrol strategy, see Stuesse 2010a, Heyman 2010). The rationale for traffic enforcement-based border control was that there were far fewer miles of road to patrol than miles of border, and that the Border Patrol could concentrate its officers at strategic transportation chokepoints north of the border, which a majority of undocumented crossers would be forced to use in order to reach destinations in the US interior.

Other factors also encouraged the traffic enforcement approach to border control. For example, the Border Patrol had been awarded an exceptionally broad authority to stop cars and ask drivers as well as passengers about their legal status. This power was rooted in a series of Supreme Court rulings in the mid-1970s which significantly reinterpreted Fourth Amendment protections against unreasonable search and seizure for drivers and passengers throughout a so-called "border region" defined by the court as a 100-mile-deep swath of land adjacent to the US–Mexico border. In the landmark case *United States v. Brignoni-Ponce*, the court allowed Border Patrol officers to use "Mexican appearance" as a key factor in deciding whether or not to stop a vehicle and ask the driver as well as passengers about their legal status. The Border Patrol had been granted this authority on account of their claim that traffic enforcement north of the US–Mexico border was the lynchpin of US border control policy and that without race-based criteria for making traffic stops—on the basis of an apparent correlation between "looking Mexican" and legal status—US border control policy would effectively collapse.

The importance of the "prevention through deterrence" strategy is hard to overestimate. As Nevins has explored in detail in his research, the shift away from interior enforcement to border enforcement per se in the mid-1990s meant that the earlier site-specific border control experiments in San Diego and El Paso quickly morphed into officially sanctioned show-of-force operations in all major urban centers along the US–Mexico border. This entailed a huge increase in the annual US border control budget, which more than doubled during the 1990s (Dixon and Gelatt 2005). One important knock-on effect of this change in policy was a "balloon effect" in which undocumented migrants were forced away from urban areas into more rugged terrain (Madsen 2007, Cornelius 2001). Indeed, by the end of the 1990s undocumented immigrants crossing the US–Mexico border were far more likely to do so at remote locations away from more traditional urban crossing points, and moreover were far less likely to attempt repeat entries during their first 12 months in the US due to the hardships associated with crossing away from more built-up areas along the border (Fernández-Kelly and Massey 2007). Another predictable outcome of the new strategy was a significant increase in border deaths due to exhaustion and dehydration, which by some accounts tripled as a result of the new policy (Eschbach, Hagan, and Rodríguez 2003, Massey, Durand, and Malone 2002, Androff and Tavassoli 2012).

Despite the significant changes in border control policy brought about by the mid-1990s "prevention through deterrence" strategy, this period of border militarization was arguably but an opening act for a more recent round of fence and wall construction at the US–Mexico border. Indeed, what happened during the 1990s at the US–Mexico border pales in comparison to the fence- and wall-building spree that happened at the border in the wake of the September 11 terrorist attacks, which, among other things, suggests that state borders as such are hardly relics of some bygone era (O'Dowd 2010, Anderson 2002).

The Border Patrol has, since 2001, opted for a risk-based approach to border control in the US Southwest. This strategy has focused on human resources, and in particular on intelligence gathering as well as the mobile deployment of special operations teams to manage challenges at the border as they develop (see, for example, US Border Patrol 2012a, US Government Accountability Office 2012). The Border Patrol's new risk-management approach in part reflects its now cautious endorsement of "prevention through deterrence." For example, there is now compelling evidence that, on the whole, the primary impact of "prevention through deterrence" has been to increase the costs and dangers of undocumented entry rather than reduce undocumented entry as such (Massey et al. 2002, Cornelius 2001). Moreover, studies suggest that the strategy actually decreased the odds of being apprehended by the Border Patrol, as a result of the increased use of professional smugglers as well as the remoteness of passable parts of the border for both undocumented migrants and Border Patrol officers (Massey and Singer 1995, Massey 2005, Spener 2009). Nonetheless, the focus on preventive infrastructure started as a result of the 1994 strategy, initially part of the glossy-format packaging of Border Patrol strategy in the US Southwest (see, for example, Andreas 1998), is perhaps more now than ever central to overall Border Patrol strategy.

Since 2005 and the launch of the so-called "Secure Border Initiative" (SBI) in the US Southwest, tactical infrastructure appropriations for fence- and wall-building have reached historic highs (Rosenblum 2013: 15–16). As a result, since 2005 the number of miles of pedestrian fences and vehicle barricades at the US–Mexico border has more than quadrupled and tripled respectively (US Government Accountability Office 2009, US General Accountability Office 2010). Some version of a border wall or fence—from simple vehicle barriers and the traditional "landing mat" fencing to comprehensive bollard-style walls and triple layer Sandia-style fences (for an inventory, see Madsen 2011)—is now virtually continuous across California, Arizona, and New Mexico, where the bulk of the now nearly 700 miles of fencing and barriers has been installed. In other words, whereas "prevention through deterrence" was mostly an urban practice during the 1990s, it now has a much broader geographical reach. It is also important to note that the ranks of the Border Patrol have grown massively since 2006 in order to deal with unfenced and unwalled sections of the border. For example, Border Patrol staffing in the US Southwest has roughly doubled since 2005, with the large majority of this increase targeted for the Texas–Mexico border where border fencing is not widely used (18,500 of 21,100 agents are assigned to the nine US Southwest Border Patrol sectors, as of 2012; 7,700 of 14,700 agents stationed in the US Southwest are based in Texas; see US Border Patrol 2012b).

The newest component of "prevention through deterrence" at the US–Mexico border is a practice that a recent comprehensive review of US Border Patrol strategy refers to as "consequences enforcement" (Meissner et al. 2012: 31–33). The goal of this new program is to reduce future undocumented entry by increasing the financial, psychological, and legal costs as well as social hardships associated with apprehension by the Border Patrol in the US–Mexico border zone. The key components of the new "consequences enforcement" approach include: the criminal prosecution of repeat undocumented entrants as felony offenders in the federal court system (Slack et al. 2013); mandatory detention for non-Mexicans caught at the border while they wait for a first immigration hearing; the expanded use of expedited removal (which allows for the extra-judicial removal of undocumented migrants caught within 100 miles of the border) (Coleman 2012a); increased use of repatriation flights to Mexico; and among other aspects, so-called "alien transfer" for first-time entrants (which allows undocumented Mexican nationals to be deported through border ports of entry far removed from where they were apprehended) (De León 2013). The major shift here is away from the practice of so-called "voluntary return"—whereby those caught by the Border Patrol in the US–Mexico border region are allowed to return to their countries of origin, usually Mexico, without legal or other consequences if they waive their right to appear before an immigration judge. In place of "voluntary return," which is now increasingly limited to humanitarian cases and/or for unaccompanied minors, the new "consequences enforcement" model makes expanded use of detention and typically enforces the maximum discretionary legal penalties associated with undocumented entry. "Voluntary return" cases have all but disappeared over the past decade, from 77 percent of all border region enforcement cases in 2005 to 14 percent in 2012. Similarly, the percentage of border region enforcement cases involving formal criminal charges, formal deportation, as well as lateral or remote repatriation jumped from 23 percent of all cases in 2005 to 86 percent of all cases in 2012 (Rosenblum 2013: 6–10).

US immigration control strategy is still very much about the practice of border control at the US–Mexico border, an important reminder that the regulation of migrant mobility is an enduring component of states' geopolitical-territorial authority. Yet, we should not assume a simple identity between border enforcement and immigration enforcement in the US case. Indeed, a geopolitical-territorial focus on the border and the problem of border control cannot come at the expense of an analysis of what we see as perhaps the most important trend in US immigration strategy since 2001: the implosion of immigration policing into the US interior as well as its dispersal across an uneven patchwork of state and local law enforcement agencies (LEAs) (Coleman 2007b, Coleman 2007a, Coleman 2009, Chavez and Provine 2009, Gilbert 2009, Golash-Boza 2013,

Heyman 2010, Hiemstra 2010, Manges Douglas and Sáenz 2013, Nuñez and Heyman 2007, Varsanyi 2008a, Varsanyi 2008b, Wells 2004, Stuesse 2010b, Walker and Leitner 2011). This development mirrors a similar de-bordering (or re-bordering) of immigration enforcement in the European context (e.g. Bigo and Guild 2005, Castañeda 2010), and in general demonstrates that everyday spaces away from state borders are increasingly important sites of immigrant surveillance and regulation by immigration authorities.

There is not the space to go into great detail about these developments, so instead we offer a quick bird's eye view of this process over the past two decades or so. The growth of US immigration enforcement in formally non-border spaces can be traced to federal laws passed in the 1990s, co-terminus with the growth in prison-building in the US, which restructured civil immigration penalties for convicted non-citizens. By the end of the 1990s lawmakers mandated detention and deportation for a range of non-citizen felony, misdemeanor, and minor infraction offenders newly defined in immigration statute in the aggregate as "aggravated felons." The result was a significant growth in the number of non-citizen administrative detainees and a spike in deportations in the late 1990s. However, much more important for the expansion of interior immigration enforcement has been the post-2001 devolution of immigration authority to non-federal law enforcement agencies. Prior to 2001, even if localities enforced the criminal provisions of the Immigration and Nationality Act (INA) (i.e. crimes triggering deportation by federal authorities), the civil aspects of the INA (i.e. relating to immigration status) were deemed off-limits to non-federal authorities (Seghetti, Viña, and Ester 2006). This is no longer the case. For example, a range of programs now allow non-federal officers to act as proxy immigration enforcement agents in the interior on behalf of federal immigration authorities. The two most important are the 287(g) and the Secure Communities programs. Although hard detention and deportation data on these programs is difficult to nail down, the data that is available suggests that since 2002, but for the most part since 2006, millions of individuals have been identified as deportable under these programs and that, of this total, several hundred-thousand individuals, at a minimum, have been formally deported (Immigration and Customs Enforcement 2013b, Rosenblum 2012, Immigration and Customs Enforcement 2013a). The 287(g) authority, available in theory since 1996 but implemented only as of 2002, allows state and local police to investigate immigration cases and make immigration arrests on behalf of federal authorities. Since 2002, 287(g) has taken two basic forms: roving operations in which police ask for immigration status during the course of routine patrolling or, more commonly, an interview procedure for suspected undocumented immigrants booked into state and/or local jails as a result of routine police work. The Secure Communities program is a variation on this second form of 287(g) policing in that it allows non-federal police to run a biometric status check on suspected undocumented immigrants booked into custody.

There are some important differences between the 287(g) and Secure Communities programs. The biggest operational difference is that the Secure

Communities program does not ensure that identified undocumented individuals will be deported; the 287(g) program, in contrast, more or less guarantees a closed custody chain between local and/or state police and federal officials. Another important difference is coverage. For example, the 287(g) program is currently operative in just 36 jurisdictions, which represents approximately a 50 percent reduction from the program's high point several years ago (Coleman 2009). In contrast, the Secure Communities program is now operational in each of the 3,144 counties in the US, mostly via sheriff's offices. But despite these differences there are some important consistencies between the two programs. We will emphasize three. First, what is important about both the 287(g) and Secure Communities programs is that they allow local authorities to hold an individual so that even when local charges run their course the individual in question is not released. Indeed, an important operational characteristic shared by 287(g) and Secure Communities is that they allow individuals to be held by the state continuously, in relation to first, criminal charges, and second, civil immigration charges, in a way that was literally unheard of 10 years ago.

A second important characteristic that links the two programs is their focus on non-criminal arrests. This is apparently contradictory in the sense that both programs now require some form of criminal suspicion and arrest on local or state charges, i.e. not on immigration grounds alone, before an immigration investigation can be started. However, being booked into custody by local and/ or state police with either 287(g) or Secure Communities authority should not be confused with conviction or criminality per se. For example, since 2006 regularly more than 50 percent of the deportee population arrested under the 287(g) program has not had any criminal charges pending when they were deported. Moreover, if we were to include other lesser offenses such as public order crimes, illegal entry, false citizenship claims, non-payment of alimony, resisting arrest, driving without a license, and such, we would see that those charged with serious crimes rarely exceeds 35 percent of the 287(g) program's total deportee population (Rosenblum 2012: 32). This trend holds also for Secure Communities, although given its nationwide operability far more individuals are affected. For example, cumulative data for 2008 through 2012 shows that less than 50 percent of the individuals deported under Secure Communities can be characterized as serious offenders. Moreover, nationwide nearly 25 percent of the program's deportees are removed under non-criminal grounds (Immigration and Customs Enforcement 2013b).

Third, and for us by far most important, 287(g) and Secure Communities partners regularly use traffic violations and traffic enforcement to stop individuals and ask about their status (Coleman 2012b). This represents a return, but in a much more spatially extensive way, to the Border Patrol's traffic-based enforcement in the US–Mexico border region prior to the "prevention through deterrence" policy, as discussed above. Indeed, it is while undocumented immigrants are on the move, between non-work spaces (shops, churches, homes, child care centers, recreational facilities, etc.) and worksites, that programs like 287(g) and

Secure Communities make their largest impact (Stuesse and Coleman 2014, Coleman and Kocher 2011). Accordingly, we characterize the focus on immigrant "automobility" under both 287(g) and Secure Communities enforcement as an "entrapment" strategy (Nuñez and Heyman 2007, Ortiz 2000, Stuesse 2010a) focused on immigrant social reproduction, and with class/race stratification as an outcome. By entrapment we mean that traffic enforcement is saturated in immigrant neighborhoods over long periods of time, deployed on feeder roads connecting significant immigrant communities to larger cross-town arteries during commuting hours, and is more concerned with alienage than motoring infractions. As a result, simple social reproduction—the unpaid work required to literally stay alive, which in most US cities today is intimately dependent on automobile use—is recast as an increasingly risky set of practices, with detention and deportation by local and/or state proxies an ever-present possibility. Gilbert refers to this sort of re-bordering away from state borders per se, we think productively, as a strategy of "incapacitation" (Gilbert 2009, see also Coutin 2010b on the increasing "inviability" of immigrant life in the US); De Genova refers to this problem slightly differently as the everyday grind of deportability for resident undocumented communities (De Genova 2002).

Our overall point in this overview is that contemporary US immigration control is constituted through two important sets of practices. On the one hand, since the mid-1990s, border control in a topographical sense has become an important part of overall US immigration control strategy, with the US–Mexico border a particular point of focus. It should be stressed that this is a recent development, which is indeed part of a global trend towards hardened geopolitical territories since the end of the Cold War (Jones 2012). On the other hand, in the wake of the militarization of the US–Mexico border, brought about by the "prevention through deterrence" doctrine, US immigration control has also become a much more general topological, or unlocatable, problem focused on everyday immigrant life in the US interior. As we hinted at briefly above, this is suggestive of earlier policing tactics by the Border Patrol in the US–Mexico border zone, but taking place at an unprecedented scale.

However, there is far more to this than simply two discrete undercurrents to US immigration control. Indeed, we propose that the geopolitical and biopolitical, or topographical and topological, aspects of US immigration control together constitute a condition that might be best referred to as "borders behind a border" (Leerkes, Leach, and Bachmeier 2012). For example, the geopolitical fortification of the US–Mexico border has a generally "hard" carceral function: in making the border much more dangerous, the "prevention through deterrence" strategy has effectively turned the US into a "zone of confinement" (Coutin 2010a) which is both difficult to enter and exit (if re-entry is intended). It is within this generally geopolitical or territorial mode of immigration control qua confinement that we think the more recent turn to interior enforcement is best theorized. Here we think the appropriate analogy is to the relationship between border control and legalization during the early 1980s, as we explored briefly in our introduction, in which amnesty was understood as dependent on border enforcement. But today this relationship is turned inside out and upside down in the sense that the very immigrant enclave communities that laws like IRCA sought to regularize are now being created by the confluence of border and interior enforcement working together. Indeed, we see the relatively new turn to interior enforcement, as benchmarked by programs like 287(g) and Secure Communities, as promoting irregularization. The 287(g) and Secure Communities programs encourage the growth of what was referred to during the lead-up to IRCA as "shadow populations"—that is undocumented communities cut off in social reproduction terms from the rest of the US, and as such largely invisibilized, even as they remain confined in the US as a crucial constituent of the contemporary American workforce. In this sense, we see the growth of "shadow populations" largely as an instance of immigrant social control, rather than, for example, a concerted attempt to reach down into the social ganglia of the US population and deport all undocumented migrants—and then keep them out via fortified geopolitical borders. Much has been made of the apparent US strategy to deport 100 percent of the resident undocumented immigrant population (see, for example, the so-called "Endgame" strategy sketched out in Department of Homeland Security 2010), but for us this is a rhetorical flourish.

What we understand is transpiring today at the crossroads of geopolitical and biopolitical immigration control is the production of a "softer" form of social, economic, and political deportation within the US interior—at least in relation to the "hard" cement and steel of the US–Mexico border. We are not saying that deportation no longer takes the classic form of territorial banishment, but that interior enforcement in the main works by using the looming threat of territorial banishment as a result of traffic enforcement and other social reproduction-specific policing, in conjunction with the specter of lethal geopolitical infrastructures like the US–Mexico border, to regulate the ways in which resident undocumented immigrant communities learn to socially reproduce as well as work. Insofar as the result is a form of deportation within which targets immigrant social reproduction, this is a paradoxical mode of deportation without the usual emphasis on physical territorial removal. In sum, the geopolitical and biopolitical designs at the heart of US immigration control come together to produce a sort of exteriorized inside rather than a simply exteriorized outside.

Conclusion: Borders are Not Everywhere

In this chapter, we have argued that topography and topology, as well as geopolitics and biopolitics, should not be read as opposed and antithetical "rule sets" for modeling how borders work, and why. We have explored this question theoretically, with reference to the split between topographical and topological research on borders, but also empirically in terms of the US–Mexico case study, which shows that US immigration control leans heavily on both geopolitical and biopolitical, or topographical and topological, borders. Moreover, the increased reliance on enforcement of topological borders in the current era depends crucially upon the continued enforcement and escalating militarization of the topographical border. In a phrase, US immigration control since at least IRCA has policed both borders and bodies, operating interdependently in the US–Mexico borderlands and beyond.

We want to conclude by cautioning against what we see as a now relatively common refrain in the humanities and social sciences literature on borders and border control: that borders are, now, everywhere. This refrain is obviously more germane to the topological approach to borders studies. Balibar, for example, notes provocatively that borders "are being thinned out and doubled, becoming border zones, regions, or countries where one can reside and live" (Balibar 2002: 92). Balibar suggests, usefully, that the "quantitative relation between border and territory is being inverted" (Balibar 2002: 92). To be clear, our reading is not that Balibar substitutes the topological for the topographic, or the biopolitical for the geopolitical. Rather, his point is to explore borders not as perimetrical limit points but also as spaces of (policed) residence. In terms of immigration control, this suggests very usefully that border patrol is a police of "things" and "people" as much as it might be a police of "edges." This is indeed why we use the phrase immigrant policing rather than immigration enforcement in our research—in order to signal that in addition to border control in a narrow topographical sense, immigration control is also about policing what immigrants do (in the interior), and how. But Balibar's provocation nonetheless risks becoming a simple "rule set" about how to see and understand borders—and indeed we have seen multiple instances of the "borders are everywhere" trope used as a somehow commonsensical shorthand to talk about borders generally at our national conferences.

The challenge here is twofold. First, we see in the "borders are everywhere" approach a tendency, no doubt unintended, to forget about the now hyper-militarization of state territorial borders. For example, we note that the topographical and topological approaches to borders are too often explained calendrically, as if the era of border militarization was an immediate post-Cold War phenomenon and that now we are on to something new. The big problem here is that a border such as the militarized interface between the US and Mexico cannot simply be folded into a general "borders are everywhere" narrative (here we are inspired by the work of Anzaldúa 1987, Lugo 2000, and Rosaldo 1989, among others). That border is very much not everywhere in the sense that it is exceptionally locationally robust and, we would add, lethal in its territorial rootedness. Indeed, this border is entirely unlike interior borders—such as traffic checkpoints in immigrant neighborhoods by sheriffs enrolled in the Secure Communities program—in the sense that it is a permanent feature of the landscape whose primary goal is to blockade and control entry, in a classical, geopolitical-territorial or topographical sense. To insist that this border be seen as one among other kinds of borders, to us, underplays what it means to cross this international boundary. Second, any generalized claim about interior borders being "everywhere" is fundamentally incorrect. If the US–Mexico border is resolutely somewhere, interior borders are, as Mezzadra and Neilson note in their work, both temporally and spatially intermittent; their overall goal is to modulate in fits and starts rather than to permanently scrutinize. This means that interior borders are sometimes not in play; they are not everywhere, but sometimes everywhere. This is not to suggest that interior borders are somehow less meaningful than geopolitical-territorial borders. Indeed, we see the intermittent and patchwork-like quality of interior borders as posing very serious challenges to resident undocumented immigrant communities in the sense that they can loom over social reproduction practices and spaces, and in this way can, as above, be described as social control governance. We also want to stress that interior borders, in their fitfulness, are far from totalizing; the capture performed by interior borders, and the constant social reproduction threat they pose, is far from complete. In particular, there are ample opportunities for creative acts by immigrant communities in the face of these borders (Nelson and Hiemstra 2008, Cravey 2003, Ridgley 2008, Lewis et al. 2013, Marrow 2009, Stuesse and Coleman 2014, Stuesse (under review), Stuesse, Grant-Thomas, and Staats (under review)), and we would add that the local and state apparatus which directs these borders is far from a coherent machinery (see, more broadly, Soguk 2007, Campbell and Heyman 2007).

References

Agnew, J.A. 1994. The Territorial Trap: The Geographical Assumptions of International Relations Theory. *Review of International Political Economy*, 1(1), 53–80.

Agnew, J.A. 1999. Mapping Political Power Beyond State Boundaries: Territory, Identity, and Movement in World Politics. *Millennium-Journal of International Studies*, 28(3), 499–521.

Agnew, J. 2008. Borders on the Mind: Re-Framing Border Thinking. *Ethics & Global Politics*, 1(4), 175–191.

Albert, M. and L. Brock. 1996. Debordering the World of States: New Spaces in International Relations. *New Political Science*, 18(1), 69–106.

Albert, M., Jacobson, D. and Y. Lapid, eds. 2001. *Identities, Borders, Orders: Rethinking International Relations Theory*. Minneapolis: University of Minnesota Press.

Allen, J. 2009. Three Spaces of Power: Territory, Networks, plus a Topological Twist in the Tale of Domination and Authority. *Journal of Power*, 2(2):197–212.

Amin, A. 2002. Spatialities of Globalisation. *Environment and Planning D: Society & Space*, 34(3), 385–399.

Amin, A. 2004. Regions Unbound: Towards a New Politics of Place. *Geografiska Annaler Series B: Human Geography*, 86(1), 33–44.

Amoore, L. 2005. Biometric Borders: Governing Mobilities in the War On Terror. Paper read at the Workshop of the British International Studies Association, May, at Aberystwyth, Wales.

Anderson, J. 2002. Borders After 11 September 2001. *Space and Polity*, 6(2), 227–232.

Andreas, P. 1998. The U.S. Immigration Control Offensive: Constructing an Image of Order on the Southwest Border. *Crossings: Mexican Immigration in Interdisciplinary Perspectives*, edited by M.M. Suárez-Orozco. Cambridge, MA: Harvard University Press, 341–361.

Andreas, P. 2000. *Border Games: Policing the US–Mexico Divide*. Ithaca: Cornell University Press.

Androff, D.K. and K.Y. Tavassoli. 2012. Deaths in the Desert: The Human Rights Crisis on the U.S.–Mexico Border. *Social Work*, 57(2), 165–173.

Anzaldúa, G. 1987. *Borderlands/La Frontera: The New Mestiza*. Second edition. San Francisco: Aunt Lute Books.

Arp, W., M.K. Dantico, and M. Zatz. 1990. The Immigration Reform and Control Act of 1986: Differential Impact on Women? *Social Justice*, 17(2), 23–39.

Balibar, É. 2002. *Politics and the Other Scene*. London: Verso.

Belcher, O., L. Martin, A. Secor, S. Simon, and T. Wilson. 2008. Everywhere and Nowhere: The Exception and the Topological Challenge to Geography. *Antipode*, 40(4), 499–503.

Bigo, D. 2001. The Möbius Ribbon of Internal and External Securities. *Identities, Borders, Orders: Rethinking International Relations Theory*, edited by M. Albert, D. Jacobson, and Y. Lapid. Minneapolis: University of Minnesota Press, 91–116.

Bigo, D. 2002. Security and Immigration: Toward a Critique of the Governmentality of Unease. *Alternatives*, 27(1), 63–92.

Bigo, D. and E. Guild, eds. 2005. *Controlling Frontiers: Free Movement Into and Within Europe*. Aldershot: Ashgate.

Brock, L. 1999. Observing Change, "Rewriting" History: A Critical Overview. *Millennium*, 28(3), 483–497.

Brunet-Jailly, E. 2007. *Borderlands: Comparing Border Security in North America and Europe*. Ottawa: University of Ottawa Press.

Calavita, K. 1992. *Inside the State*. New York: Routledge.

Calavita, K. 1996. The New Politics of Immigration: "Balanced Budget Conservatism" and the Symbolism of Proposition 187. *Social Problems*, 43(3), 284–305.

Campbell, H. and J.M. Heyman. 2007. Slantwise: Beyond Domination and Resistance on the Border. *Journal of Contemporary Ethnography*, 36(1), 3–30.

Castañeda, H. 2010. Deportation Deferred: "Illegality," Visibility, and Recognition in Contemporary Germany. *The Deportation Regime: Sovereignty, Space, and the Freedom of Movement*, edited by N. De Genova and N. Peutz. Durham: Duke University Press.

Chavez, J.M. and D.M. Provine. 2009. Race and the Response of State Legislatures to Unauthorized Immigrants. *Annals of the American Academy of Political and Social Science*, 623, 78–92.

Chock, P.P. 1995. Ambiguity in Policy Discourse: Congressional Talk about Immigration. *Policy Sciences*, 28(2), 165–185.

Coleman, M. 2007a. A Geopolitics of Engagement: Neoliberalism, the War on Terrorism, and the Reconfiguration of US Immigration Enforcement. *Geopolitics*, 12(4), 607–634.

Coleman, M. 2007b. Immigration Geopolitics Beyond the Mexico–US Border. *Antipode*, 38(2), 54–76.

Coleman, M. 2008. Between Public and Foreign Policy: US Immigration Law Reform and the Undocumented Migrant. *Urban Geography*, 29(1), 4–28.

Coleman, M. 2009. What Counts as the Politics and Practice of Security, and Where? Devolution and Immigrant Insecurity after 9/11. *Annals of the Association of American Geographers*, 99(5), 904–913.

Coleman, M. 2011. Topologies of Practice. *Dialogues in Human Geography*, 1(3), 308–311.

Coleman, M. 2012a. Immigrant Il-legality: Geopolitical and Legal Borders in the US, 1882–present. *Geopolitics*, 17(2), 402–422.

Coleman, M. 2012. The "Local" Migration State: The Site-Specific Devolution of Immigration Enforcement in the U.S. South. *Law & Policy*, 34(2), 159–190.

Coleman, M. and A. Kocher. 2011. Detention, Deportation, Devolution and Immigrant Incapacitation in the U.S., Post 9/11. *Geographical Journal*, 177(3), 228–237.

Connolly, W.E. 1974. Essentially Contested Concepts in Politics. *The Terms of Political Discourse*. Oxford: Blackwell, 9–44.

Cornelius, W.A. 1998. Appearances and Realities: Controlling Illegal Immigration in the United States. *Temporary Workers or Future Citizens? Japanese and US Migration Policies*, edited by M. Weiner and T. Hanami. New York: New York University Press, 384–428.

Cornelius, W.A. 2001. Death at the Border: Efficacy and Unintended Consequences of US Immigration Control Policy. *Population and Development Review*, 27(4), 661–685.

Coutin, S. 2010a. Confined Within: National Territories as Zones of Confinement. *Political Geography*, 29(2), 200–208.

Coutin, S. 2010b. Exiled by Law: Deportation and the Inviability of Life. *The Deportation Regime: Sovereignty, Space, and the Freedom of Movement*, edited by N.P. De Genova. Durham: Duke University Press, 351–370.

Cravey, A.J. 2003. Toque una ranchera, por favor. *Antipode*, 35(3), 603–621.

Dávila, A., J.A. Pagán, and M.V. Grau. 1999. Immigration Reform, the INS, and the Distribution of Interior and Border Enforcement Resources. *Public Choice*, 99(3–4), 327–345.

De Genova, N.P. 2002. Migrant "Illegality" and Deportability in Everyday Life. *Annual Review of Anthropology*, 31(1), 419–447.

De León, J. 2013. The Efficacy and Impact of the Alien Transfer Exit Programme: Migrant Perspectives from Nogales, Sonora, Mexico. *International Migration*, 51(2), 10–23.

Department of Homeland Security. 2010. *Endgame: Office of Detention and Removal Strategic Plan, 2003–2012*. Washington, DC: Department of Homeland Security.

Diener, A.C. and J. Hagen, eds. 2010. *Borderlines and Borderlands*. Lanham: Rowan and Littlefield.

Dillon, M. 2007. Governing Terror: The State of Emergency of Biopolitical Emergence. *International Political Sociology*, 1(1), 7–28.

Dixon, D. and J. Gelatt. 2005. Immigration Enforcement Spending Since IRCA. *MPI Immigration Facts*. Washington, DC: Migration Policy Institute.

Dodge, M. and R. Kitchin. 2005. Code and the Transduction of Space. *Annals of the Association of American Geographers*, 95(1), 162–180.

Donnan, H. and T.M. Wilson, eds. 2010. *Borderlands: Ethnographic Approaches to Security, Power, and Identity*. Lanham: University Press of America.

Dunn, T.J. 1996. *The Militarization of the U.S.–Mexico Border, 1978–1992: Low-Intensity Conflict Doctrine Comes Home*. Austin: Center for Mexican American Studies Books, University of Texas at Austin.

Dunn, T.J. 2010. *Blockading the Border*. Austin: University of Texas Press.

Elden, S. 2007. Govemmentality, Calculation, Territory. *Environment and Planning D: Society & Space*, 25(3), 562–580.

Elden, S. 2011. What's Shifting? *Dialogues in Human Geography*, 1(3), 304–307.

Eschbach, K., J. Hagan, and N. Rodríguez. 2003. Deaths during Undocumented Migration: Trends and Policy Implications in the New Era of Homeland Security. *In Defense of the Alien*, 26, 37–52.

Fall, J.J. 2010. Artificial States? On the Enduring Geographical Myth of Natural Borders. *Political Geography*, 29(3), 140–147.

Fernández-Kelly, P. and D.S. Massey. 2007. Borders for Whom? The Role of NAFTA in Mexico–US Migration. *Annals of the American Academy of Political and Social Science*, 610, 98–118.

Flores, R.R. 2002. *Remembering the Alamo: Memory, Modernity, and the Master Symbol*. Austin: Center for Mexican American Studies, University of Texas Press.

Gilbert, L. 2009. Immigration as Local Politics: Re-bordering Immigration and Multiculturalism through Deterrence and Incapacitation. *International Journal of Urban and Regional Research*, 33(1), 26–42.

Golash-Boza, T. 2013. Mapping the Shift from Border to Interior Enforcement of Immigration Laws during the Obama Presidency. *Social Scientists on Immigration Policy*. Available at: http://stopdeportationsnow.blogspot.com/2013/01/mapping-shift-from-border-to-interior_7232.html. Accessed: 9/10/2013.

Gonzalez Baker, S. 1997. The "Amnesty" Aftermath: Current Policy Issues Stemming from the Legalization Programs of the 1986 Immigration Reform and Control Act. *International Migration Review*, 31(1), 5–27.

Gottmann, J. 1973. *The Significance of Territory*. Charlottesville: University Press of Virginia.

Heyman, J.M. 2010. The State and Mobile People at the US–Mexico Border. *Class and Contention in a World in Motion*, edited by W. Len and P. Gardiner Barber. New York: Berghahn Press, 58–78.

Hiemstra, N. 2010. Immigrant "Illegality" as Neoliberal Governmentality in Leadville, Colorado. *Antipode*, 42(1), 74–102.

Huysmans, J. 2000. The European Union and the Securitization of Migration. *Journal of Common Market Studies*, 38(5), 751–777.

Immigration and Customs Enforcement. 2013a. *Fact Sheet: Delegation of Immigration Authority Section 287(g) Immigration and Nationality Act* [cited September 6, 2013].

Immigration and Customs Enforcement. 2013b. Secure Communities: Nationwide Interoperability by Conviction Report (Monthly Statistics through May 31, 2013). Washington, DC: Department of Homeland Security [online]. Available at: http://www.ice.gov/doclib/foia/sc-stats/nationwide_interop_stats -fy2013-to-date.pdf.

Johnson, C., R. Jones, A. Paasi, L. Amoore, A. Mountz, M. Salter, and C. Rumford. 2011. Interventions on Rethinking "The Border" in Border Studies. *Political Geography*, 30(2), 61–69.

Jones, R. 2009. Categories, Borders and Boundaries. *Progress in Human Geography*, 33(2), 174–189.

Jones, R. 2012. *Border Walls: Security and the War on Terror in the United States, India and Israel*. London; New York: Zed Books.

Katz, C. 2001. On the Grounds of Globalization: A Topography for Feminist Political Engagement. *Signs*, 26(4), 1213–1234.

Kolossov, V. 2005. Border Studies: Changing Perspectives and Theoretical Approaches. *Geopolitics*, 10(4), 606–632.

Kuus, M. 2010. Critical Geopolitics. *The International Studies Encyclopedia*, edited by R.A. Denemark. Available at: http://www.isacompendium.com/ subscriber/tocnode?id=g9781444336597_yr2011_chunk_g97814443365975_ ss1-28. London: Blackwell.

Lapid, Y. 2001. Identities, Borders, Orders: Nudging International Relations Theory in a New Direction. *Identities, Borders, Orders: Rethinking International Relations Theory*, edited by M. Albert, D. Jacobson, and Y. Lapid. Minneapolis: University of Minneosta Press, 1–20.

Latham, A. 2011. Topologies and the Multiplicities of Space-Time. *Dialogues in Human Geography*, 1(3), 312–315.

Latour, B. 2005. *Reassembling the Social: An Introduction to Actor Network Theory*. Oxford: Oxford University Press.

Leerkes, A., M. Leach, and J. Bachmeier. 2012. Borders Behind the Border: An Exploration of State-Level Differences in Migration Control and their Effects on US Migration Patterns. *Journal of Ethnic and Migration Studies*, 38(1), 111–129.

Lewis, P.G., D.M. Provine, M.W. Varsanyi, and S.H. Decker. 2013. Why Do (Some) City Police Departments Enforce Federal Immigration Law? Political,

Demographic, and Organizational Influences on Local Choices. *Journal of Public Administration Research and Theory*, 23(1), 1–25.

Limón, J.E. 1994. *Dancing with the Devil: Society and Cultural Poetics in Mexican–American South Texas*. Madison: The University of Wisconsin Press.

Lugo, A. 2000. Theorizing Border Inspections. *Cultural Dynamics*, 12(3), 353–373.

Madsen, K.D. 2007. Local Impacts of the Balloon Effect of Border Law Enforcement. *Geopolitics*, 12(2), 280–298.

Madsen, K.D. 2011. Barriers of the U.S.–Mexico Border as Landscapes of Domestic Political Compromise. *Cultural Geographies*, 18(4), 547–556.

Manges Douglas, K. and R. Sáenz. 2013. The Criminalization of Immigrants and the Immigration-Industrial Complex. *Daedalus, the Journal of the American Academy of Arts and Sciences*, 142(3), 199–227.

Marrow, H.B. 2009. Immigrant Bureaucratic Incorporation: The Dual Roles of Professional Missions and Government Policies. *American Sociological Review*, 74(5), 756–776.

Marston, S.A., J.P. Jones, and K. Woodward. 2005. Human Geography without Scale. *Transactions of the Institute of British Geographers*, 30(4), 416–432.

Martin, P.L. 1982. Select Commission Suggests Changes in Immigration Policy: A Review Essay. *Monthly Labor Review*, 105(2), 31–39.

Martin, P.M. 2005. Comparative Topographies of Neoliberalism in Mexico. *Environment and Planning A*, 37(3), 203–220.

Massey, D. 1991. The Political Place of Localities Studies. *Environment and Planning A*, 23(2), 267–281.

Massey, D.S. 2005. Backfire at the Border. *CATO Institute Center for Trade Policy Studies*, 29, 1–16.

Massey, D.S., J. Durand, and N.J. Malone. 2002. *Beyond Smoke and Mirrors: Mexican Immigration in an Era of Economic Integration*. New York: Russel Sage Foundation.

Massey, D.S. and A. Singer. 1995. New Estimates of Undocumented Migration and the Probability of Apprehension. *Demography*, 32(2), 203–213.

Meissner, D., D.M. Kerwin, M.A. Chishti, and C. Bergeron. 2012. *Immigration Enforcement in the United States: The Rise of a Formidable Machinery*. Washington, DC: Migration Policy Institute.

Meyers, D.W. 2005. US Border Enforcement: From Horseback to Hi-Tech. *MPI Insight*. Washington, DC: Migration Policy Institute.

Mezzadra, S. and B. Neilson. 2003. Né qui, né altrove—Migration, Detention, Desertion: A Dialogue. *Borderlands e-journal*, 2(1). Available at: http://www.borderlands.net.au/vol2no1_2003/mezzadra_neilson.html.

Mezzadra, S. and B. Neilson. 2008. Border as Method, or, the Multiplication of Labor. *Transversal*. Available at: http://eipcp.net/transversal/0608/mezzadraneilson/en.

Mezzadra, S. and B. Neilson. 2013. *Border as Method, or the Multiplication of Labor*. Durham: Duke University Press.

Nagar, R., V. Lawson, L. McDowell, and S. Hanson. 2002. Locating Globalization: Feminist (Re)readings of the Subjects and Spaces of Globalization. *Economic Geography*, 78(3), 257–284.

Nelson, L. and N. Hiemstra. 2008. Latino Immigrants and the Renegotiation of Place and Belonging in Small Town America. *Social & Cultural Geography*, 9(3), 319–342.

Nevins, J. 2002. *Operation Gatekeeper: The Rise of the Illegal Alien and the Making of the US–Mexico Boundary*. London: Routledge.

Newman, D. 2002. From "Moribund Backwater" to "Thriving into the Next Century": Political Geography at the Turn of the New Millennium. *The Razor's Edge: International Boundaries and Political Geography*, edited by C. Schofield, D. Newman, A. Drysdale, and J. Allison Brown. London: Kluwer Law International, 3–19.

Newman, D. 2003. On Borders and Power: A Theoretical Framework. *Journal of Borderland Studies*, 18(1), 13–25.

Newman, D. 2006. Borders and Bordering: Towards an Interdisciplinary Dialogue. *European Journal of Social Theory*, 9(2), 171–186.

Newman, D. and A. Paasi. 1998. Fences and Neighbors in the Postmodern World: Boundary Narratives in Political Geography. *Progress in Human Geography*, 22(2), 186–207.

Nuñez, G. and J.M. Heyman. 2007. Entrapment Processes and Immigrant Communities in a Time of Heightened Border Vigilance. *Human Organization*, 66(4), 354–365.

O'Dowd, L. 2010. From a "Borderless World" to a "World of Borders": Bringing History Back In. *Environment and Planning D: Society & Space*, 28(6), 1031–1050.

Paasi, A. 1996. *Territories, Boundaries, and Consciousness: The Changing Geographies of the Russian–Finnish Border*. London: John Wiley & Sons.

Paasi, A. 2005. Generations and the "Development" of Border Studies. *Geopolitics*, 10(4), 663–671.

Paasi, A. 2011. Geography, Space and the Re-Emergence of Topological Thinking. *Dialogues in Human Geography*, 1(3), 299–303.

Paredes, A. 1958. *With His Pistol in His Hand: A Border Ballad and its Hero*. Austin and London: University of Texas Press.

Parker, N. and N. Vaughan-Williams. 2009. Lines in the Sand? Towards an Agenda for Critical Border Studies. *Geopolitics*, 14(3), 582–587.

Pérez, E. 1999. *The Decolonial Imaginary: Writing Chicanas into History*. Bloomington: Indiana University Press.

Purcell, M. and J. Nevins. 2005. Pushing the Boundary: State Restructuring, State Theory, and the Case of US–Mexico Border Enforcement in the 1990s. *Political Geography*, 24(2), 211–235.

Ridgley, J. 2008. Cities of Refuge: Immigration Enforcement, Police and the Insurgent Genealogies of Citizenship in US Sanctuary Cities. *Urban Geography*, 29(1), 53–77.

Rosaldo, R. 1989. *Culture and Truth: The Remaking of Social Analysis*. Boston: Beacon Press.

Rosas, G. 2012. *Barrio Libre: Criminalizing States and Delinquent Refusals of the New Frontier*. Durham: Duke University Press.

Rosenblum, M.R. 2012. *Interior Immigration Enforcement: Programs Targeting Criminal Aliens*. Washington, DC: Congressional Research Service.

Rosenblum, M.R. 2013. *Border Security: Immigration Enforcement between Ports of Entry*. Washington, DC: Congressional Research Service.

Rosière, S. and R. Jones. 2012. Teichopolitics: Re-considering Globalisation through the Role of Walls and Fences. *Geopolitics*, 17(1), 217–234.

Rumford, C. 2006. Theorizing Borders. *European Journal of Social Theory*, 9(2), 155–169.

Saldívar, J.D. 1997. *Border Matters: Remapping American Cultural Studies*. Berkeley: University of California Press.

Seghetti, L.M., S.R. Viña, and K. Ester. 2006. Enforcing Immigration Law: The Role of the State and Local Enforcement. *CRS Report for Congress*. Washington, DC: Congressional Research Service, Library of Congress.

Slack, J., D.E. Martinez, S. Whiteford, and E. Peiffer. 2013. In the Shadow of the Wall: Family Separation, Immigration Enforcement and Security. Preliminary Data from the Migrant Border Crossing Study. Tucson: Center for Latin American Studies, University of Arizona.

Soguk, N. 2007. Border's Capture: Insurrectional Politics, Border-Crossing Humans, and the New Political. *Borderscapes: Hidden Geographies and Politics at Territory's Edge*, edited by P.K. Rajaram and C. Grundy-Warr. Minneapolis: University of Minnesota Press, 283–308.

Spener, D. 2009. *Clandestine Crossings: Migrants and Coyotes on the Texas-Mexico Border*. Ithaca: Cornell University Press.

Staeheli, L.A. and C.R. Nagel. 2006. Topographies of Home And Citizenship: Arab-American Activists in the United States. *Environment and Planning A*, 38(9), 1599–1614.

Stuesse, A. and M. Coleman. 2014. Automobility, Immobility, Altermobility: Surviving and Resisting the Intensification of Immigrant Policing. *City and Society*, 26(1).

Stuesse, A., A. Grant-Thomas, and C. Staats. Under Review. As Others Pluck Fruit Off the Tree of Opportunity: Immigration, Racial Hierarchies, and Intergroup Relations Efforts in the United States. *Du Bois Review*.

Stuesse, A. 2010a. Challenging the Border Patrol, Human Rights and Persistent Inequalities: An Ethnography of Struggle in South Texas. *Latino Studies*, 8(1), 23–47.

Stuesse, A. 2010b. What's "Justice and Dignity" Got to Do with It? Migrant Vulnerability, Corporate Complicity, and the State. *Human Organization*, 69(1), 19–30.

Stuesse, A. Under Review. *Globalization Southern Style: Immigration, Race, and Work in the U.S. South*. University of California Press.

Thrift, N. 2008. *Non-representational Theory: Space, Politics, Affect.* London: Routledge.

Trujillo, M.L. 2009. *Land of Disenchantment: Latina/o Identities and Transformations in Northern New Mexico.* Albuquerque: University of New Mexico Press.

Turner, B.S. 2007. The Enclave Society: Towards a Sociology of Immobility. *European Journal of Social Theory,* 10(2), 287–304.

US Border Patrol. 2012a. 2012–2016 Border Patrol Strategic Plan. Washington, DC: US Customs and Border Patrol Protection.

US Border Patrol. 2012b. US Border Patrol Sector Profile Fiscal Year 2012, ed. Available at: http://www.cbp.gov/linkhandler/cgov/border_security/bord er_patrol/usbp_statistics/usbp_fy12_stats/usbp_sector_profile.ctt/usbp_ sector_profile.pdf. Washington, DC: US Border Patrol.

US Government Accountability Office. 1995. *Border Control: Revised Strategy is Showing Some Positive Results.* Washington, DC: Government Accountability Office.

US Government Accountability Office. 2009. *Immigration Enforcement: Controls over Program Authorizing State and Local Enforcement of Federal Laws Should be Strengthened.* Washington, DC: US Government Accountability Office.

US General Accountability Office. 2010. Secure Border Initiative: DHS Has Faced Challenges Deploying Technology and Fencing Along the Southwest Border. Washington, DC: General Accounting Office.

US Government Accountability Office. 2012. *Border Patrol Strategy: Progress and Challenges in Implementation and Assessment Efforts.* Washington, DC: Government Accountability Office.

US House of Representatives Judiciary Committee. 1986. House Report (Judiciary Committee) No. 99–682 (I). Washington, DC.

US Select Commission on Immigration and Refugee Policy. 1981a. US Immigration Policy and the National Interest: Staff Report of the Select Commission on Immigration and Refugee Policy. Washington, DC.

US Select Commission on Immigration and Refugee Policy. 1981b. US Immigration Policy and the National Interest: The Final Report and Recommendations of the Select Commission on Immigration and Refugee Policy to the Congress and the President of the United States. Washington, DC.

Varsanyi, M.W. 2008a. Immigration Policing through the Backdoor: City Ordinances, the "Right to the City," and the Exclusion of Undocumented Day Laborers. *Urban Geography,* 29(1), 29–52.

Varsanyi, M.W. 2008b. Rescaling the "Alien," Rescaling Personhood: Neoliberalism, Immigration, and the State. *Annals of the Association of American Geographers,* 98(4), 877–896.

Vaughan-Williams, N. 2009. The Generalised Bio-political Border? Re-conceptualising the Limits of Sovereign Power. *Review of International Studies,* 35(4), 729–749.

Vélez-Ibañez, C. 1996. *Border Visions: Mexican Cultures of the Southwest United States.* Tucson: University of Arizona Press.

Walker, K.E. and H. Leitner. 2011. The Variegated Landscape of Local Immigration Policies in the United States. *Urban Geography*, 32(2), 156–178.

Walters, W. 2002. Deportation, Expulsion, and the International Police of Aliens. *Citizenship Studies*, 6(3), 265–292.

Walters, W. 2006a. Border/Control. *European Journal of Social Theory*, 9(2), 187–203.

Walters, W. 2006b. Rethinking Borders Beyond the State. *Comparative European Politics*, 4(2/3), 141–159.

Wells, M.J. 2004. The Grassroots Reconfiguration of US Immigration Policy. *International Migration Review*, 38(4), 1308–1347.

Wilson, T.M. and H. Donnan. 2012. Borders and Border Studies. *A Companion to Border Studies*. Oxford: Wiley-Blackwell, 1–25.

Woodward, K., John Paul Jones III, and Sallie A. Marston. 2010. The Eagle and the Flies, a Fable for the Micro. *Area*, 42(3), 271–280.

Zolberg, A.R. 2006. *A Nation by Design*. Cambridge: Harvard University Press.

SECTION II
Border Work by Non-Traditional Actors Near the Border

Chapter 4

Locating the Border in Boundary Bay: Non-Point Pollution, Contaminated Shellfish, and Transboundary Governance

Emma S. Norman

Introduction

Locating the border in everyday life is, perhaps, slightly easier if you are perched on its edge. The people that live around the waters of Boundary Bay—an inlet of the Strait of Georgia between British Columbia and Washington—are reminded daily of the tensions between the "borderlessness" of nature (particularly water) and the processes and acts of "bordering." Looking out from its shores, the beautiful waters of Boundary Bay may appear indistinct from other bays in North America. However, the difference becomes clear when those living along the Bay try to visit with their neighbors on the opposite side. Travelling the short distance requires a commitment in time and resources to cross from one sovereign state, Canada, into another, the United States. The uncertainty of long borderlines and the increased police scrutiny at the checkpoints makes the once easy jaunt across the border increasingly difficult—particularly post-9/11.

In this chapter, I explore how one group attempts to transcend the divisions of the international border through ongoing collaboration. The Shared Waters Alliance (SWA) is a transboundary, multi-stakeholder group that was organized to reduce pollution inputs in Boundary Bay (see Figure 4.1 for a map and Table 4.1 for a list of participants). The analysis of this regional, yet transboundary, environmental organization grapples with questions regarding scale, borders, and governance. Similar to other chapters in this volume, the narrative complicates the connection between borders and the sovereign state.

The analysis provides the opportunity to continue the engagement with Balibar's concept of the "everyday border" (2004). In this chapter, I engage with this theme by querying how actors are able to "locate the border in everyday life." In the Boundary Bay context, I examine the extent to which local actors are able to participate actively, engage, and affect the governance of transboundary waters. The case of Boundary Bay is an important contribution to this volume because it explores how a connected ecosystem can be politically fragmented and the consequences of that fragmentation in terms of water quality and environmental governance. In addition, the case highlights how citizen actors are able to align

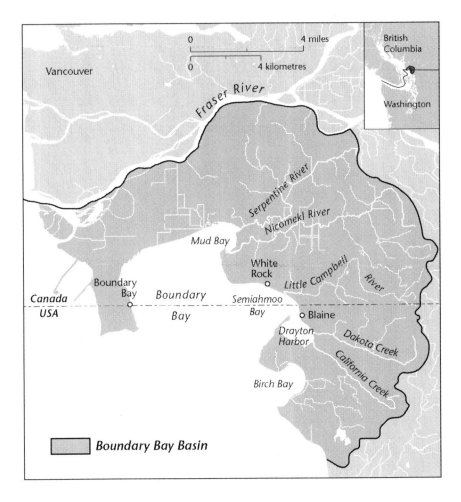

Figure 4.1   Map of Boundary Bay Basin, British Columbia (Canada) and Washington (United States)

Source: Original map. Cartography by Eric Leinberger, University of British Columbia.

themselves through place (the Bay), rather than through stratified jurisdictions to meet shared goals.

Drawing on empirical evidence from interviews and participant observation,[1] I analyze how the members of the Shared Waters Alliance mobilize to govern water across the international border. To aid in this analysis, I explore the specific

---

1   This is part of a wider project that looked at changing transboundary water governance patterns along the Canada–US border. Overall, 75 interviews were conducted between 2006 and 2009. Approximately 10 of those were affiliated with Boundary Bay.

techniques that the SWA employs to govern transboundary water. I also show how the SWA's environmental education campaigns utilize both a nature–society approach and a cultural, economic approach to promote their cause. Analyzing these specific techniques helps to engage in concepts of social construction of borders and borderland identities. Thereby, I explore how the SWA constructs a discourse of "borderless" or "boundless" nature in terms of "nature" and "resource use," while experiencing a very "bordered" environment, organizationally.

Background

In the Northwestern continental United States and Southwestern Canada lies a small body of water, aptly named Boundary Bay. Historically, Boundary Bay was one of the most productive shellfish harvesting locations on the Pacific coast. Indigenous communities relied successfully on these waters for centuries as primary sources of food. Between 1920 and 1950, the bay provided upwards of 50 percent of annual commercial oyster harvesting in British Columbia in the early years of commercial harvesting (MacKenzie 1996). However, degraded upland environments, bacterial contamination, and excess fecal coliform prompted governmental officials to close the area for harvesting in 1962.[2] In Washington, the bay only recently opened for restricted use; it remains closed in British Columbia.

Political fragmentation and the divided managing authorities for the Bay made identifying the source of the contamination challenging. Initially, the closures of the shellfish beds led to finger pointing across the border; the dirty water was assumed to be the *other* country's responsibility. Americans blamed the Canadians and the Canadians blamed the Americans for the polluted waters. When asked about the accountability of the pollution, David, a Canadian stakeholder involved in the environmental stewardship of Boundary Bay reflected in an interview:

> At the beginning, for example, Canadians really believed that the sewage system on the US side was *soooo* bad that it had to be coming over to the Canadian side. And the Americans—I mean it was just kitchen talk on the US side—well the Canadians are so polluted and the Little Campbell is so polluted, it's got to be impacting what is happening.

Water quality studies indicated, however, that the primary sources of the contamination were from sources on both sides of the border and included: increased population pressures, agricultural runoff, increased urbanization, and a faulty sewage system in Blaine, Washington (Cheung 2003, Hay and Co 2003). In addition, a survey of outfalls along the Bay (Cheung 2003) indicated that the Little

---

2   Fecal coliform bacteria is found in the digestive tracts of warm-blooded animals and is used as an indicator of pathogens such as the virus hepatitis and the bacteria E. coli. If ingested, it could be a potentially serious human health risk (Picot et al. 2011).

Campbell River (LCR) in British Columbia was a significant contributor of fecal contamination to Boundary Bay.

In response to the water pollution and the subsequent closure of the shellfish beds, a group called the Shared Waters Roundtable mobilized in 1999. This group, renamed the Shared Waters Alliance (SWA) in 2005, is part of the surge of interest in regional, binational coordination (Norman and Bakker 2009). Initially, this group consisted of a few civil servants from British Columbia.[3] However, as word spread about the on-the-ground efforts to deal with the pollution inputs in the bay, the group grew to include indigenous representatives, local stakeholders, civil servants, and environmental organizations from both the province of British Columbia and state of Washington. The focus of the group has also broadened to reflect wider issues such as storm water runoff and community outreach.[4] Overall, there are approximately 45 people listed as members of the roundtable, although on average about 12 people representing various levels of government, Environmental Non-governmental Organizations (ENGOs), or citizen groups attend meeting. Table 4.1 lists the members that are, or have been, part of the SWA:

Table 4.1   Organizations and agencies currently or historically represented at Shared Waters Alliance

Governments Agency (Federal/ Provincial)	Governments Agency (Regional/ Local)	Indigenous groups	Private	Education	Non-profit/ ENGO
Environment Canada	Metro Canada	Semiahmoo First Nation	Hirsch Consulting	University College of the Fraser Valley	Delta Farmland and Wildlife Trust
Fisheries and Oceans Canada	Township of Langely	Nooksack Tribe	Tynehead Hatchery	Kwantlen Polytechnical Institute	A Rocha
US Environmental Protection Agency	City of Surrey		Bay Reality		BIEAP/ FREMP
BC Ministry of the Environment	Whatcom County		Delta Corp		Nicomekl Enhancement Society
BC Ministry of Agriculture, Food and Fisheries	Whatcom Conservation District				

---

3   The founding members represented Environment Canada, Fraser Valley Health and City of Surrey.

4   As the focus of the group changes, the membership base of the group is also changing (i.e. less involvement of private industry as the focus moves away from the closure and into outreach).

The SWA is a multi-stakeholder organization working towards the general health of the Boundary Bay ecosystem. Representatives include state and non-state actors from both Canada and the US (including federal,[5] provincial, state, and municipal employees as well as local, NGO, public, and private stakeholders). As a provincial employee and member of the SWA commented in an interview, "The experiences of the SWA reflect how a multi-jurisdictional watershed with numerous non-point sources of pollution can be managed through a coordinated effort among different stakeholders." The SWA (2007) describe their primary goal as:

> To meet shellfish harvesting standards through working with other jurisdictions, whether dealing with water quality, education or changes in practice initiatives.

To meet this goal, they outline the following objectives:

1. To characterize and identify key sources of contamination to Boundary Bay; and
2. To undertake outreach and pollution prevention projects which reduce contamination levels in tributaries and the Bay itself.

One of the initial contributions of the group is the facilitation of the *Boundary Bay Circulation Study*, which helped identify the point sources of pollution for the marine system. The study suggested that the greatest benefit to increasing the water quality would be to reduce fecal coliform levels in the Little Campbell River, which feeds into Semiahmoo Bay, and Dakota Creek, which feeds into Drayton Harbor.

Participating members of the SWA are all volunteers—there are no paid staff members or central office. Rather, the members try to streamline their workload to participate in this project and coordinate grants to run specific projects (such as the creation of the *Boundary Bay Circulation Study* and the *Boundary Bay Habitat Atlas*) and in-kind contributions (such as alliance members' time and use of meeting rooms). The majority of the active participants of the SWA live within the Boundary Bay basin or—to a lesser degree—in the nearby metropolitan areas of Vancouver and Seattle. The group holds bimonthly meetings designed to encourage information exchange and coordination of stewardship and educational activities.

The creation of the SWA occurred at a time when the political climate has been amenable to the creation of regional, subnational transboundary groups. The development was part of a wider shift related to shifting patterns of water governance along the Canada–US border (see Norman and Bakker 2009). In the Pacific Coast context, the signing of the British Columbia–Washington Environmental Cooperation Agreement in 1992, and later the formation of the Puget Sound Action Team in 1996 and the Georgia Basin Ecosystem Initiative in 1998, provided the political momentum for the development of a group such as the SWA. The

---

5   At the regional level (i.e. Seattle-based and Vancouver-based agencies).

founding members, in effect, were riding a wave of interest (and subsequent funding) in regional, transboundary governance. This approach departs from the government–government approach found in organizations such as the International Joint Commission (IJC) and represents a wider trend in environmental governance where multi-jurisdictional actors, at a subnational scale, are increasingly engaged in transboundary governance activities (Norman, Cohen, and Bakker 2013).

Shared Waters Alliance and the "Local Trap"

The increase of local, non-governmental actors in transboundary water issues reflects a broad shift from "government" to "governance" of natural resources (Rhodes 1996; Herod et al. 1998; Swyngedouw 2000a, 2000b; Gibbs et al. 2001; Jessop 2003, 2004). This change in governance patterns applies to a wide range of water issues, from the protection of marine habitat to equitable access to and distribution of water. For the latter, international organizations such the World Water Council (WWC) consider local authorities as critical to the fulfilment of community water needs through the provision of water services and access to water and sanitation. However, limited institutional capacity and few federal–local partnerships continue to impede the impact of local governance activities (WWC 2007). Some proponents of governance deploy the concept to assert the importance and/or necessity of involving actors at local scales. Others express doubt about the concept's usefulness, for example querying the conceptual limits to scalar concepts (such as "the local scale") as a means of inquiry into governance issues (Brenner 2001; Marston, Jones, and Woodward 2005; Jonas 2006; Norman, Bakker, and Cook 2012).

Some scholars, however, have challenged the uncritical acceptance of the rhetoric of the local in environmental policy (Evans 2004). This fits with Brown and Purcell's (2005) description of the "local trap," which assumes that organization, policies, and action at the local scale are inherently more likely to have desired social and ecological effects than activities organized at other scales.[6] This raises the risk of treating the involvement of local actors in water management in a relatively uncritical fashion, particularly with respect to assumptions of equitable and meaningful participation, significant influence over decision-making, and accountability or capacity (Van Rooy 1997, 2004, Taylor 2004).

The SWA case provides insight into the role of local actors in regional, transboundary water governance. Specifically, the case explores the role of local actors as either "active participants" and "agents of change," as scholars such as Ostrom (1990) suggest, or as political pawns used for political cachet, which scholars such as Cochrane (1986) suggest.

---

6   This concept, although originally applied to political ecology, is transferable to the work within environmental governance, as both have limited engagement with politics of scale literature.

Within this context, I discuss techniques employed by the SWA to govern transboundary water. Specifically, I analyze their governance structure and their environmental education programs.

Bordered Spaces and Asymmetrical Governance

On the surface, the SWA looks like a truly balanced binational committee. However, after closer investigation, it becomes apparent that the physical make-up of the committee remains decidedly asymmetrical (and "bounded" by state). For example, although the group's mandate is binational, the meetings are primarily held in British Columbia (mostly Surrey), leading to significantly more participants from Canada than from the US. The official roster has an equal number of participants listed; however, in a sampling of six meetings only one member from the US attended (with an average participation rate of 12 people). Not surprisingly, when the meetings are held in the US, the number of US participants increases. In order to help mitigate this asymmetry, the group leaders recently agreed to meet *at* the border. Although the change in venue provided a neutral setting, minimized the amount of travel for the US participants, and eliminated the need to cross through customs, the asymmetry continued: only one US participant attended. This asymmetrical participation of actors exemplifies wider trends reported by transboundary governance actors, where inconvenient meeting venues created a barrier to fluid borders and a limiting factor for civic engagement.

Furthermore, internal boundaries within the committee contributed to divisive politics. Although the discourse describing the ecosystem omits the political borders, national divides remain pronounced in the imagination of the actors. When discussion turns to governing the shared ecosystem, national biases and "us and them" terms often prevail, particularly as the SWA committee attempted to find the cause of the pollution in Boundary Bay. Before the circulation study, which generally showed how (and where) the pollution was distributed, participants consistently looked to one another to blame. However, after the study found a relatively even distribution of pollution, the national biases were slightly tempered and the groups could focus their energy on working collaboratively. For example, the group worked on creating a *Boundary Waters Atlas*, coordinated an annual waterway cleanup, and worked to influence decision-makers to support projects that focus on improving water quality of the Bay.

The reification of borders also occurred beyond the nation-state level. Members made clear distinctions between members of the SWA. These internal boundaries served to position people within the governance process. The most pronounced included governmental and non-governmental; however, other boundaries were also present, including indigenous–non-indigenous and federal–local. The collaborative process in some ways attempts to create an even playing field for the actors. However, this process shows that erasing the distinctions between groups of people is difficult (if not impossible) even in these binational, multi-jurisdictional fora, which strive to create neutral spaces for negotiation and collaboration.

In the section below, I illustrate how the SWA attempts to imagine (and enact) a region based on hydrological rather than political borders. Matthew Sparke's conceptualization of Cascadia helps with this comparison.

Social Construction of Landscape: Cascadia, the Salish Sea, and the SWA

This faltering construction of a borderless bioregion is reminiscent of the Cascadia project, often employed by environmental movements as a symbol for nature-based politics. Sparke (2002b: 12) defines Cascadia as:

> [A] concept cross-border region, an idealized transnational space on the Pacific coast of North America, bridging the 49th parallel and linking the Canadian Province of British Columbia and the United States of Washington and Oregon.

Like the SWA, Cascadia has failed to mobilize in the general population's imagination as a singular region despite two decades of promotion by environmental advocates and, more recently, business elite. Interestingly, proponents of Cascadia still tout the region as an emerging transitional region-state (and a distinct cross-border region) despite its lack of economic and political integration and the lack of tangible outcomes prophesized by its ardent supporters. Part of the drive is the powerful metaphor based on images of a singular, beautiful bioregion. This idea has also emerged with the new term, "Salish Sea," which in 2010 became the official name to refer to the waters between the Strait of Juan de Fuca and Haro Strait. The name had been used for more than two decades by environmentalists, indigenous groups, and scholars, however the official naming was deemed an important step to publicly recognize the pre-colonial boundaries of the region and to begin to conceive the region as a connected ecosystem (Tucker 2013).

The popularization of the Ernest Callenbach's (1975) book *Ecotopia* partly contributes to the creation of the concept of Cascadia, and arguably, the Salish Sea as a regional identity. The widespread readership of *Ecotopia* helped create an imagined geography that portrayed the Pacific Northwest as a united (and uniform) geographic area whose residents were environmentally-minded. Joel Garreau's 1980s bestseller the *Nine Nations of North America* also puts forward the idea of Ecotopia—stretching from Santa Barbara, California to Southeast Alaska—in which a shared conservationist ethic and acceptance of alternative lifestyles fostered a regional society which transcended national differences (Coates 2002). These conceptualizations of the Pacific Northwest as a unified geographic continuum, whose residents, arguably, share a strong environmental ethic helped lay the groundwork for the *idea* of a shared binational bioregion. This ideology transformed into an idiom in the late 1980s when environmentalists started characterizing the coastal Pacific as "Cascadia."

The idea of Cascadia aims to be emotive, a geographic imagining that helps to conjure up images of a "natural landscape." Cascadia, as Sparke (2002b) notes, is a long-running project based on continually constructing images of a shared geography. The branding of Cascadia as a trans-border bioregion is based on the physical mountain range that runs north–south between British Columbia and Washington, spanning the Canada–US border (Alper 1996, 1997). Although rooted in bioregional idealism, the notion of Cascadia has grown to include neoliberal entrepreneurial projects.

In fact, Sparke contends that the geoeconomic aims of the region have actually replaced the earlier environmentally based geopolitical aims. He suggests that conservative business forces and politicians who view the bioregion as a way to build economic strongholds now replace the environmental Cascadia of Callenbach's imagining. As with the SWA's positioning of the shellfish industry (discussed below), this reinvention is part of wider trends, in which "a new vocabulary of geoeconomics has replaced the old vocabulary of geopolitics" (Sparke 2002a: 216).

Thus, despite the constructed discourse of "borderless" or "boundless" nature in terms of "nature" and "resource use," both the Cascadia and the SWA projects experience a very "bordered" environment. Within this context, I explore the variable techniques the SWA employs to govern transboundary water at a local scale. In particular, I explore how the relative fluidity and fixity of the border influence the actors' ability to govern the shared waters.

Local Representation in Multi-Jurisdictional Committees

As discussed above, the environmental governance literature tends to position the local actor in a privileged position without a critical engagement with their actual capacity to govern. Looking closely at the SWA, it becomes clear that while local actors have many positive contributions, their decision-making capacity is decidedly limited.

Local NGO actors provide a number of valuable contributions such as the role of environmental educator and advocate. Those that work within the SWA, report that the local ENGO's played a pivotal role in the organizational structure of the group through an assumed role of group liaison. For example, the local actors help to provide momentum for the committee and help to keep the meetings running. They also relay information between governmental and non-governmental sectors.

As groups often have the tendency to work within "institutional silos," these multi-stakeholder forums help to widen the discussion and perpetuate practices of inclusive governance. Several respondents involved in environmental governance noted the importance of information sharing in their work. This is particularly true in transboundary cases, where asymmetrical governance structures between Canada and the US governments (as in other border geographies) benefit from neutral platforms to help "level" the playing field.

Within the SWA committee, the local organizations tend to function as a connector of people, ideas, and issues. For example, Margaret Cuthbert, one of the founding members of the SWA described her role as a "switchboard operator" where she and her husband, David, were "basically conducting traffic" between groups. When there was an institutional lull in the committee, she was able to help prod the governmental employees into assuming different roles. Margaret reflected in an interview at her home in White Rock, British Columbia on the early stages of the committee and her role in getting the group organized.

> [At that point] there were no action items—no leadership, nothing. At that meeting, I went over to [one member of the committee] because she is a real go-getter. If we can get a chair for this committee, would you be willing to donate the minutes. She said 'yes.' And then, I went over to the province and I took [the committee member] aside—because I knew his skills—and I said 'would you be willing to chair the meeting if someone else took the minutes' and he said, 'I will think about it.' And, so, I got them to talking to each other. And then ... there was suddenly a change in the committee ... and the whole dynamic changed. And everyone now knew that we would have *real* minutes. Before ... everyone was quite frustrated and yet [the governmental agencies] weren't prepared to do it.

Here, Margaret drew on personal experience and connections to foster leadership within the committee. Knowledge of the participants' personalities and their professional capacity (i.e. whether they were "action-oriented," driven by this specific issue, or had the time to invest in committee work) helped in the coordination of a new committee structure. Orchestrating the logistical fabric of the SWA—such as who would be the note-taker, who could provide meeting venues, and who could chair the meetings—proved a huge asset. The initiative, local knowledge, and passion for the issues enhance the capacity of the committee.

Leadership also plays a significant role in the governance process and helps lesson the impacts of a "bordered" environment. Passionate (and often out-spoken) local stakeholders provide momentum for the maintenance of the committee. For example, one state employee involved in the SWA described Geoff—Founder of Drayton Harbor Community Shellfish—as an "enviro-warrior" who brought unfaltering determination to the issue of water quality. Geoff was described as "determined, driven and a true leader" in the environmental field. The state employee reflected in an interview:

> With watershed planning, you always need leadership and you need spark plugs
> gave both like few I have ever seen. And he is still going.

Another role of citizen actors in the SWA members is to maintain momentum with committee work. The local stakeholders assume the role of advocates for the environment by instigating action (and warding off complacency) in the governance process. They also propel the process by asking questions and probing governmental employees to take their action further. As Margaret noted:

> I think that the real impetus [of the SWA] has been from the grassroots. I think that the agencies went along because there was an opportunity through the Georgia Basin Ecosystem Initiative. Then, after the five years the Georgia Basin Action Plan, there was a motivating factor in terms of funding that gave staff in Canada something to work with and they wanted to *appear* to be doing something. But the *real* impetus came from [stakeholders] ... And then when we came along—and we were just *mouthy*—we were the non-profit kind-of trying to figure out what the heck's going on, mouthy volunteers. And slowly and surely we got to figure out a lot more of what was going on. And I think that our role has not changed that much. I think we still are the thorn that prods action and I think that's a valid description.

Here, Margaret describes the role of citizen actors as "thorn that prods action." She suggests that by being visible or "mouthy" they were able to solicit action within the group. Conversely, she suggests that mandates and the need to "appear" to meet public expectations largely drive the work of civil servants. A range of respondents (both governmental and non-governmental actors) made this distinction between "mandated work" and "principled work." For example, one Washington state employee said:

> I wish that the government entities could be more like [the environmental groups]—passionate and dedicated. However, the government employees are often more conservative—more timid.

Sustaining energy in the group, therefore, proved to be a key role for citizen actors. The ability for a local group to focus on a specific project (rather than directed by multiple and sometimes conflicting mandates) helped the group to maintain focus. This role was particularly important as governmental employees reported that over-extended schedules, demanding workloads, and a shifting political climate tended to limit their participation in the stakeholder process.[7]

A tangible example of the local contribution of the SWA is the habitat-mapping project, which ultimately led to the production of the *Boundary Bay Habitat*

---

7 This is not to say that the characterization of governmental employees as "passive" should be extended across the board. Many of the governmental actors interviewed across the Canada–US border have similar characterizations of those described for the NGOs (passionate and dedicated) and have made significant gains in transboundary water governance approaches. However, scale and authorship tend to temper individual laureates in governmental cases. For example, within the governmental sector individual authorship is less common (i.e. your work is *for* the Province, the State, the City, or the Nation, writ large, rather than a more specific—smaller scale—local group).

Atlas. The multi-level connections served to bring volunteers together to map the shoreline using Global Positioning System devices and then enter the information into digital databases. The local groups organized the mapping project by drawing on a large volunteer base—in both Washington and British Columbia—through community outreach campaigns and personal contacts. In the end, funding and expertise from various levels of governance—both governmental and non-governmental—and in both Canada and the US helped to complete the project.

The ability of local actors to remain active participants in multi-jurisdictional and international issues is, in large part, a matter of timing and the type of issue addressed. As Van Rooy (1997: 92) finds in her work on civil society, NGOs tend to have the most influence on "highly salient, low policy issues accompanied by early and continuous access to decision-makers." This matches the conditions of the SWA Boundary Base case, where there is great public interest in the issue, but it remains a "low-policy issue" because the issue lacks crisis status (ibid.). This situation tends to foster greater involvement at the subnational level, as crises—at the international level—are likely to attract the attention of federal-level employees. The involvement of local actors in the early stages of the SWA development contributed to benefits associated with a vertically integrated governance structure.

Information Sharing and Access to Policy Makers

Information sharing is another benefit of these regional binational groups. SWA respondents consistently reported this as one of the greatest strengths of the group. Through the roundtable process, actors gleaned information from different sectors and different jurisdictions. This information exchange occurs through a variety of mediums, including: bimonthly meetings, digital listserves, co-investigation of studies, co-production of reports, field trips, and shared work parties. For example, the SWA meeting minutes document this type of multi-jurisdictional exchange:

> Environment Canada is in discussions with the Semiahmoo First Nation regarding water quality in Semiahmoo Bay and Little Campbell River. Through the Beach Hero program, we realize that poaching is a big issue, and we are collecting data on poaching. There is a new regulation in place now that prohibits discharge of sewage from recreational boaters. There is a possibility of Environment Canada being able to fund outreach activities through the Georgia Basin Action Plan. Please contact AC or JB for more information. (October 4, 2006)

Another example of the multi-jurisdictional exchange:

> IJC is working with 4 other Semiahmoo band members to argue against the new treaties that Gordon Campbell is attempting to establish. They are also working toward claim for access to marine animals as part of aboriginal right for sustenance, and also pursuing avenues to facilitate the cleanup of the water quality in the estuary so that it may be re-opened for the band/public.

> MC voiced concerns regarding the dog off-leash concerns in the East White Rock beach area. JB noted that she recently received a phone call from a citizen regarding the issue, and that signage in the area may help with foreshore education. (August 2, 2006)

In the first case, the Semiahmoo First Nation representative shared information regarding their work on treaty-rights as well as access to marine mammals. Grounding these complex issues of treaty-rights to the site of Boundary Bay helps other members of the SWA connect with the issues and increase the scope of its relevance beyond the immediate parties. In the second case, the need for educational signage regarding off-leash dogs makes its way from a concerned citizen to the committee. The item, flagged for action, goes through the appropriate jurisdictional channels. Whether this comes to fruition, is largely dependent on the priority (and funding possibilities) of the responsible office. However, documenting the concern helps to initiate this process of action.

Another example of information sharing appears in one of the early meeting minutes (1999):

> Washington State Department of Ecology accompanied Environment Canada and BC MOE staff on a tour of the Little Campbell River watershed in May. The tour was a good opportunity to discuss the methodology used in Washington State for watershed analysis and similar challenges on both sides of the border of Boundary Bay. The Department of Ecology has produced a document outlining their approach to watershed analysis and a case study of Drayton Harbor and is now undertaking a case study of Birch Bay.

In this example, the meetings and information exchange of actors helped spur parallel research studies in Canada and the United States. However, as discussed below, the coordination of field trips has markedly declined in recent years.

Mitigating Asymmetry

The stakeholders' forum also helps to neutralize (although not completely) the political asymmetries of the actors. In this process, the ENGO group reported that access to the governmental employees was a tremendous asset to their cause. As one ENGO member reflected:

> What we find at those meetings, by sharing and having the dialogue, they hear from the grassroots things that they would not hear otherwise. When they are speaking with each other—they are speaking from a certain perspective, that is what they are allowed to talk about within a certain mandate and they have to be careful because of the current political climate. And then they hear something from the grassroots … they hear something from the community that is insisting on what needs to be done—it reminds them of the things that are not getting

done. They then have the opportunity to take that back, which I think performs a function which is a really nice opportunity—a nice opportunity for them to really get the goods and it is a really good opportunity for the community to be inputting.

In the above passage, the local community members view direct access to policy makers as a way to advance the community voice into decision-making processes. This is consistent with Van Rooy's (2004) findings that access to policy makers was an essential component of building institutional capacity for the NGOs. Dialogue with different scales of government allows the transfer of ideas from "the ground up." The multi-stakeholder format opens up space for a dialogue between different stakeholder groups and agencies that would not necessarily be in conversation otherwise. This mixing of groups helps to stave off the silo effect, in which the decision-making process omits the consideration of other sectors.[8] In this case, the local groups have political agency—the access to the governmental decision-makers gives them a sense of power. This agency counters the "local trap" concept in which local groups tend to be present in name only.

Access to governmental employees was particularly salient in a transboundary setting, as local groups received unprecedented access to policymakers from across the border. The respondents considered this important because the groups were not as familiar with the political structure or governance processes of the "other country." Having a forum where the officials and stakeholders from both sides of the border were present helped the members mitigate the asymmetry of governance structure and reduce the perceived "impenetrable" bureaucracy of an unfamiliar government process.

One group indicated that they, "perhaps naively," thought that the presence of US governmental officials in the SWA could provide some political advantage for greater action on the Canadian side. They reflected:

Maybe the Americans … maybe if it is proven that there is something that comes from the Little Campbell and crosses the border—maybe the Americans can say something to the Canadians that can get us motivated. But, it really has not been proven in any way, shape, or form and it has become obvious to us that neighbors at this level—don't make those demands of each other.

The local NGO stakeholders had envisioned transboundary work to provide more opportunities to place political pressures at a federal level. This group was disappointed to find regional governance (i.e. "neighbors at this level") had limited access to "higher-level" decision-making power (i.e. a direct line to Ottawa or Washington, DC). More specifically, they were frustrated that US delegates did not have the capacity (or at least they did not employ that capacity) to pressure the Canadian government to act more resolutely on the cleanup of Boundary Bay.

---

8   This "silo effect" continually came up in the interviews and the transboundary workshop as a barrier to transboundary cooperation and local engagement in governance issues.

This point challenges the governance literature, which often champions local participation as the most appropriate scale to govern water. However, rarely is it postulated that the local might engage in this process to have better access to the federal or supranational scale of governance. This is salient particularly within a transboundary setting, where the watershed is simultaneously "local" and "international" and the managers transect across several political jurisdictions.

Although the SWA members did not always realize the full advantages of a transboundary forum in terms of politicking, information sharing produced tangible benefits. For example, the three most common barriers to transboundary cooperation reported by respondents include: mismatched governance structures, different governance cultures, and different mandates. Providing a forum where actors meet on a regular basis and in relatively informal manner provides a great opportunity to remove these barriers. In general, the meetings foster an information exchange about current issues (environmental, policy, economic, social) facing Boundary Bay. For example, every meeting starts with an educational presentation on environmental or policy issues surrounding Boundary Bay. These presentations serve multiple purposes. First, it provides the members with up-to-date information regarding the scientific and policy efforts that are occurring throughout the region. Second, it provides a chance for a specific actor to discuss his/her individual role, as well as the role of his/her office, in protecting the Boundary Bay habitat. This information provides a way to navigate the bordered geographies and complexities of multi-jurisdictional governance. Even those that are not present at the meeting receive the meeting minutes in a digital Listserve, which often includes a copy of the presentation and the presenter's contact information. This governance structure fosters an information exchange that contributes to closing the knowledge-gap between transboundary actors.

Another asset of the SWA meetings is the dialogue that occurs between members. The group discussion contributes more than just exchange of facts and information; it also helps to create a shared vocabulary around the issues. The SWA members represent a wide range of disciplines, such as: civil servant, tribal representative, water technician, municipal manager, water engineer, and environmental education volunteer. Each of these participants comes to the table with a distinct professional culture, which is rooted in its own nuanced language and worldview. Several of the SWA members that I interviewed suggested that having a multi-disciplinary, multi-jurisdictional organization helps members break down these institutional silos. These reflexive opportunities occur as actors meet outside of their everyday institutional framework—e.g. while at multi-stakeholder meetings, workshops, or conferences. The quote below from an ENGO representative highlights the importance of this interchange:

> You know, we were sitting around, and I think I mentioned this to you, that one of the provincial folks—I think she was from up North. She said, 'well, we dialogue with the local community' and when I asked her 'who?'—she named these groups that are, you know, at such a remote level—I said, 'well that is *not*

the grassroots, that is *not* the local community—those are interests that are so
far removed from the real community—how did you ever get real community
input?' 'Oh,' she said, 'I never thought of it that way.'

In this case, the provincial employee was referring to working with a non-profit
organization that operates at a province-wide scale, rather than at a local scale.
In the discussion, the provincial employee's conflation of "non-governmental"
with "local" troubled the ENGO representative. The interchanging of "local,"
"regional," or "watershed" scales by both the practitioners and theoreticians of
environmental governance also illustrates the ambiguity between jurisdiction and
scale (Paquerot 2007) and the social construction of scale itself (Marston, Jones,
and Woodward 2005).

   One way to mitigate this ambiguity is through dialogue. The governmental and
non-governmental actors were able to negotiate a common understanding of the
term "local" through dialogue, thereby decoupling the assumed local scale with
non-profit status. This reflexive process provided space for a new understanding
of a common term, which held very different meanings to each of the participants.
Through collaborative governance, the different sectors are able to negotiate a
shared discourse, an essential step for project coordination.

   The SWA helps to create a neutral setting, where actors representing multiple
scales and multiple professions, can come to the table to discuss issues of shared
concern. This governance process follows the Integrated Watershed Management
(IWM) approach and watershed approach where hydrological borders seek to
replace political borders for the purpose of management.

   In this section, I illustrated how local actors contribute to the transboundary
governance process. In the next section, I turn to the specific techniques that
the SWA use to enact environmental governance. I then present the limitations
(barriers) to governance—particularly in a post-9/11 context.

Environmental Education: Giving Voice to Dirty Water

A common role for local ENGOs is to give voice to, and raise awareness of, ecological
systems (Van Rooy 1997; Tvedt 2004). Normalizing the concept of "saving" the
environment achieves this heightened awareness by employing techniques of eco-
governmentality[9] and ethopolitics.[10] For Boundary Bay, this voice is enacted both

---

   9   The idea of ecological governmentality builds on Foucault's idea of biopolitics;
pushing the concept forward by applying it to non-human forms. For an in-depth discussion
of eco-governmentality, see Darier (1999).

   10   Rose (1999) defines ethopolitics as a way "to characterize the ways which these
features of human individual and collective existence—sentiments, values, beliefs—have
come to provide the 'medium' within which the self-governing of autonomous individual
can be connected up with the imperatives of good government" (477).

within the wider community, through environmental education campaigns, and within the stakeholders group meetings as environmental advocates.

ENGOs often employ environmental education campaigns to impress upon the public the importance of protecting their communities' ecosystem. The SWA uses two methods in their education campaign for clean water: a nature society approach and a cultural, economic approach. The former champions the protection of the environment for "nature's sake," while resource protection, specifically shellfish harvesting, drives the latter. The latter conflates economic and environment terminology, thus folding resource economy into the natural world.

SWA members such as Geoff Menzies take the position that protecting water quality helps to revitalize a local resource economy. However, unlike other more public/visible resources (i.e. forestry), part of the education effort requires normalizing the idea that shellfish harvesting is "worthy" of saving. Because shellfish harvesting is often outside of the experience (and discourse) of many of the residents, it is sometimes difficult to obtain wide support. In the SWA case, the environmental education programs attempt to contextualize the shellfish by using the (in)ability to harvest as an indicator of the general health of the ecosystem. As one member of the SWA reflected:

> We like to talk with people about how the shellfish themselves are an indicator. You can't eat the shellfish, then your water is dirty. People might say, no, no, as long as you can swim in it it's clean—but that is like saying next year, as long as you can wash your feet in it, it is clean.

Here, the stakeholder voices concern that people's expectations for clean water (and expectations for a viable shellfish population) are continually degraded. The bay then suffers from continuous neglect as the population views shellfish harvesting as remnant of the past.

This is particularly challenging in a politically bifurcated watershed, because the policies towards shellfish beds and water quality are notably different on the US and Canadian sides of the watershed. In Canada, when Boundary Bay closed in 1962, the Canadian government in essence turned its back on the problem. The government closed the shellfish to harvesting; therefore, it became a non-issue. In the US, however, different political mandates to protect marine waters drove an effort to address the shellfish closure. The different governmental approaches to the problem perpetuate different cultural attitudes (see Norman 2013 for full discussion of this). The work of the SWA helps neutralize the asymmetrical governance structures—bringing the parties into conversation with each other.

The education campaigns, then, serve multiple purposes: to modify behavior to improve water quality standards, maintain a (political–economic) culture of shellfish harvesting, and construct the waterscape as shared (beyond the political boundary). However, keeping up with the education campaigns proves trying particularly due to the rapid growth rate in the area. Many of the people moving to the Boundary Bay region have no historical connection to the place or nostalgia for shellfish harvesting (as a resource commodity or cultural practice). As one environmental stakeholder noted:

> I would say at least half of the population of White Rock here now never lived in White Rock at a time when you could do recreational harvest [pre-1962], let alone there was commercial harvest—so they can't even relate to it as an issue. So, you talk with these people, and they say, well … it's like talking to people about having fish this big that you could catch in a river. Well, that's the past—There is no point in talking about this, so they are not really interested and then, when we take it a step further, and, yeah, but it is a total perfect indicator of general health, so, if you are willing to let it go, you are willing to let everything go. Is that what you really want to do? Then maybe people think about it a little bit.

In an effort to raise awareness of the water quality issues in Boundary Bay, one environmental group uses a community shellfish farm to engage the public and attempt to construct a shared ecological bioregion. This effort, as one state employee reflected, has a way of bringing people together and saying "you know we are all in this together." By making shellfish farming a public event—and creating an embodied experience through the physical act of harvesting—the community of participants is more likely to support measures to protect its habitat (Taylor 2004; Van Rooy 2004). Their investment of time and experience translates into an investment of concern for the wider ecosystem. One state official advocates for the community shellfish:

> Let's do the work, let's keep it open, let's sell the product, let's reinvest the money in the community and the projects and make it a really positive cycle of success.

In this case, the respondent employs political economic terminology to advocate for the maintenance of clean water and the protection of a resource industry. Conflating economic and environment terminology commonly occurs in environmental governance literature. Using terms such as "productive" and "competitive" to describe ecological systems has appeared in environmental literature since Darwin's *Origin of Species*. This collapse of disciplinary discourse is unsurprising considering the over-arching and far-reaching influence of the Industrial Revolution. In Darwin's case, his marriage into one the largest industrialist families in England, certainly had an impact on how he framed his reflections on nature (Worster 1977). Some suggest that the conflation of natural and economic terminology is part of a wider neoliberal approach to environmental issues (Sparke 2002a). The caption under the EPA's shellfish education page supports this claim:

> Protecting shellfish as a culture, a community investment and an economic sector is a joy and responsibility that belongs to all of us as community members, shoreline property owners, businesses and government agencies.

This trend is observable in other resource industries as well. Mansfield (2003a, 2003b), for example, draws on the imitation crab industry to show how biophysical discourses are folded into economic geographies. This process occurs "by making distinctions about the world that then become important for how economic activity can work" (2003a: 176). Similarly, Bridge and McManus (2000) explore how the resource sectors of mining and forestry adopt discourses of sustainability in order to avoid crises of environmental criticism. They argue, "regulation of the forestry and mineral sectors in contemporary market economies is increasingly achieved through the deployment and co-optation of narratives of sustainability" (2000: 11).

For Boundary Bay, the discussions of governing (protecting) the water for "nature's sake" or for the revitalization of a resource economy based on shellfish harvesting are increasingly inter-mixed. This is unsurprising considering one of the main community environmental education programs for the area was born from the remnants of a private shellfish business, as discussed in the next section.

The Business of Saving Shellfish

In 1992, an eager entrepreneur and trained agriculturist purchased a shellfish farm on Boundary Bay (specifically Drayton Harbor). The new owner was ready to make the business investment a life-long pursuit. However, three years after the purchase, Washington State closed down several shellfish sites due to excess bacterial contamination, including Drayton Harbor. Thus, the new owner closed the business prematurely.

In reaction to the closure, a small cohort of local actors committed the next 10 years to cleaning the water. Through a lot of hard work and multi-jurisdictional coordination between city, NGO, private, and tribal actors, a small portion of the bay (on the US side) reopened for shellfish harvesting with conditional permitting (excluding rain events, from which the water quality significantly declines). Although the opened area was too small to sustain a business operation, it attracted the attention of a non-profit environmental group interested in using the site as an educational model.

The ENGO enlisted the help of the previous business owner, who had the knowledge, facilities, and passion for shellfish harvesting, to turn the operation into a community education site. This site, which is still in operation today, aims to bring shellfish harvesting back into the fabric of the community by creating widespread socio-economic–political buy in. They publicize their work through weekly appearances at the local farmers' market and through occasional write-ups in the regional newspapers. Several of the actors involved in the clean-up efforts now participate in the SWA.

Environmental Education Along the Bay

The stakeholders on the Canadian side of the bay participate in a more traditional model of environmental education which champions the environment—as "nature"

—through programs such as Ocean Days, Stream Keepers, and Birds on the Bay. They distinguish their interests as a "matter of principle," rather than "mandates that are of their own interest." Because the actors do not personally benefit (i.e. "we don't live, we don't swim, we don't harvest" on the bay) they see themselves as enacting a voice for the ecosystem. As one Canadian member of the SWA reflected:

> We come at it as a point of principle. To us, the water should be clean. It is that simple. It should be clean, it should be swimmable. It should be clean for the critters' sake. Let alone if we want to eat them or not.

However, as they engage in dialogue with other groups and individuals involved in Boundary Bay stewardship, they also recognize the cultural and economic importance of shellfish harvesting. Both of these approaches, the nature society and the cultural–economic, attempt to modify community behavior for the same result—fostering a natural environment clean enough to sustain a harvestable shellfish population.

Although both of the above-mentioned approaches situate their cause within a wider transboundary landscape, the specific activities rarely span the Canada–US border. The Drayton Harbor shellfish project is less than a mile from the border; however, the volunteers primarily are US-based. Similarly, the education projects based in the greater White Rock region rarely capture volunteers from beyond the border, despite framing their project with hydrological rather than political borders. Occasions do exist where the White Rock and Blaine communities coordinate projects (such as the digital mapping project described above).

However, this sharing of activities has reportedly declined due to increased security measures at the border. A recent study of transportation trends in North American ports of entry post-9/11 details these changes (Olmedo and Soden 2005). They estimate that border wait times have increased from an average of 20 to 30 minutes to two to three hours. These delays reflect the material consequences of the increased security for living along, and/or frequently crossing, the Canada–US border. They also show, more generally, how the Canada–US border has become less "passenger friendly" post-9/11, revealing a shift of national priorities from expediting entry to securing borders. A 2007 article in the *Globe and Mail* (Nov. 14, 2007) suggests that the increased security is straining diplomatic relationships between Canada and the US. The headline reads, "U.S. security turning border into parking lot, envoy warns." More recently, a comprehensive study by the Fraser Institute finds that "security first" orientation within many branches of the US government have led to a thickening of the border, which has damaged the economic relationship between Canada and the US (Moens and Galber 2012).

For those involved in transboundary activities the increased border wait times have material consequences. As one resource manager in British Columbia reflects:

> Post-9/11 the infrastructure to continue coordinating binationally just began to decline.

Another SWA member from Washington State shared the sentiment:

> After 9/11, the borderlines are making it more and more difficult to justify attending these binational meetings—it ends up being a whole day.

Stakeholders are increasingly reluctant to commit to meetings that require border crossing with conditions described repeatedly as "difficult" and "unfriendly." The long delays and increased security policies influence volunteers groups and NGOs who are less likely to get financial compensation for their time traveling. The increased security also affects the volunteer base and coordination of activities.

Respondents also reported a tempering of coordination because border guards increasingly interpret old border policies with greater stringency. As another SWA member reflects:

> We used to hold water-sampling parties in Canada and the US and coordinate joint reports. However, post-9/11 it is very difficult to do this. On the US side, they say that we are 'stealing volunteer positions from Americans' ... even though this was on the books in the past, it was never enforced. Now, it is very difficult to coordinate projects across the border.

Even the physical transferring of educational material from one border town to the next has become tenuous. As one respondent reported, "Quick jaunts to the tourism office have become half day ordeals." Thus, the SWA respondents suggest that increased border security influences the institutional capacity of these regional, transboundary groups. The tendency of the SWA to hold their meetings primarily in Canada, rather than switching venues between countries, aggravates these issues.

Given the increased border wait times, it is curious that the SWA has not employed more virtual communications to augment its meetings. Although, they have Listserves, the group could really benefit from other technologies (such as Adobe Connect or Google Hangout) to facilitate virtual groups meetings. Since the Beijing Conference in 1995, virtual tools increasingly support the collaboration of long distance (Smith 2003) and intra-state (Weerasinghe 2004) networks with noted success. The possibilities to increase this collaboration using online tools such as Second Life (Polischuk 2007, Sawyer 2007) and gaming and digital simulation programs are noteworthy (Gerson 2007). The fact that the SWA has, to date, failed to employ these tools raises questions regarding the ability for the SWA to maintain cohesion (and a transboundary bioregion identity) without the physical act of meeting.

The isolated activities and the noted impacts of increased security post 9/11 suggest that although the SWA members may view the Boundary Bay region as a porous and connected border region, the border remains fixed in terms of practical operations. It also suggests that the border bounds local actors more than their governmental counterparts, contrary to the environmental governance literature, which often depicts the local as more flexible.

Conclusion

In this chapter, I used the SWA to consider how the border is located in everyday life, as it relates to transboundary ecosystem management. The SWA example highlights the potential of local actors participating in transnational governance activities. The passion and local knowledge of the local actors create a dynamic institutional body, which may strengthen multi-jurisdictional governance mechanisms. However, inadequate institutional capacity and divergent political systems limit the participation of local actors in the SWA case. In particular, the ability for the actors to coordinate projects and the binational cohesion—more broadly—has come under increasing strain post-9/11. As discussed in this chapter, as in other chapters in this volume—particularly Lamb's and Miggelbrink's chapters, working within an international setting greatly accentuates the barriers to achieving goals of shared governance.

This chapter shows how the SWA members interact with the border in attempts to transgress it for ecosystem conservation. Similar to other chapters in this volume, the narrative complicates the connection between borders and the sovereign state. Although the shared vision of maintaining a viable shellfish industry drove several of the actors to navigate through bordered spaces, the political force of the border ultimately thwarted the members attempt to create a "blurred" bioregion with hydrological borders. This analysis provides two main insights into the governance of transboundary water. One, the way the SWA presents itself (and its causes) in the education campaigns are indicative of how they perceive and construct transboundary space. Secondly, the design (and outcomes) of their projects provides insight into the material impacts of the border.

In the next chapter, Madsen continues this discussion through an engagement with actors along the US–Mexico border. Although the geopolitical and sociopolitical situation is different from the Canada–US border context, similarities of tensions at the border is found, especially within the scope of the individual, the body. Ultimately, it is the transgressions of these geopolitical borders that provide interesting—and at times surprising—similarities across contexts.

References

Alper, D.K. 1996. The Idea of Cascadia: Emergent Transborder Regionalisms in the Pacific Northwest-Western Canada. *Journal of Borderlands Studies*, 10(2), 1–22.

Alper, D.K. 1997. Transboundary Environmental Relations in British Columbia and the Pacific Northwest. *American Review of Canadian Studies*, 27(3), 359–83.

Bridge, G. and McManus, P. 2000. Sticks and Stones: Environmental Narratives and Discursive Regulation in the Forestry and Mining Sectors. *Antipode*, 32(1), 10–47.

Brown, C.J. and Purcell, M. 2005. There's Nothing Inherent about Scale: Political Ecology, the Local Trap, and the Politics of Development in the Brazilian Amazon. *Geoforum*, 36(5), 607–24.

Callenbach, E. 1975. *Ecotopia*. Berkeley: Banyan Tree Books.

Cheung, A. 2003. Semiahmoo Bay Water Quality Project. Prepared for Environment Canada.

Darier, E. 1998. *Discourses of the Environment*. Malden, Penn: Blackwell.

Evans, J. 2004. What is Local about Local Environmental Governance? Observations from the Local Biodiversity Action Planning Process. *Area*, 36(3), 270–79.

Garreau, J. 1980. *Nine Nations of North America*. Boston: Houghton Mifflin.

Gerson, S. 2007. *Possibilities for Market-based Environmental Governance via Second Life*. Washington, DC: Woodrow Wilson International Center for Scholars.

Gibbs, D.C., Jonas, A.E.G., Reimer, S., and Spooner, D.J. 2001. Governance, Institutional Capacity and Partnerships in Local Economic Development: Theoretical Issues and Empirical Evidence from the Umber Sub-region. *Transactions of the Institute of British Geographers*, 26(1), 103–19.

Hay & Co. Consultants. 2003. Semiahmoo Bay Circulation Study—Technical Report. Prepared for Environment Canada. 20pp. + Appendices.

Herod, A., O'Tuathail, G., and Roberts, S.M. 1998. *An Unruly World: Globalization, Governance, and Geography*. London: Routledge.

Jessop, B. 2004. Hollowing out the "Nation-state" and Multilevel Governance. *A Handbook of Comparative Social Policy*, edited by P. Kennett. Cheltenham, UK: Edward Elgar Publishing, 11–25.

Jessop, B. 2003. The Limits to Limits to Capital. *Antipode*, 36(3), 480–96.

MacKenzie, C. 1996. History of Oystering in the United States and Canada, Featuring the Eight Greatest Oyster Estuaries. *Marine Fisheries Review*, 58(4), 1–78.

Mansfield, B. 2003a. From Catfish to Organic Fish: Making Distinctions about Nature as Cultural Economic Practice. *Geoforum*, 34(3), 29–42.

Mansfield, B. 2003b. "Imitation Crab" and the Material Culture of Commodity Production. *Cultural Geographies*, 10(2), 176–95.

Marston, S.A., Jones, J.P., and Woodward, K. 2005. Human Geography without Scale. *Transactions of the Institute of British Geographers*, 30(4), 416–32.

Moens, A. and Gabler, N. 2012. *Measuring the Costs of the Canada–US Border*. Calgary: Fraser Institute, 45. Available at: www.fraserinstitute.org.

Norman, E. 2013. Who's Counting? Spatial Politics, Ecocolonisation, and the Politics of Calculation in Boundary Bay. *Area*, 45(2), 179–87.

Norman, E.S. and Bakker, K. 2009. Transgressing Scales: Transboundary Water Governance across the Canada–U.S. Border. *Annals of the Association of American Geographers*, 99(1), 99–117.

Norman, E.S., Cohen, A., and Bakker, K. 2013. *Water without Borders? Canada, the United States and Shared Waters*. Toronto: University of Toronto Press.

Norman, E.S., Bakker, K., and Cook, C. 2012. Introduction to Themed Section: Water Governance and the Politics of Scale. *Water Alternatives*, 5(1), 52–61.

Ostrom, E. 1990. *Governing the Commons: The Evolution of Institutions of Collective Action*. Cambridge: Cambridge University Press.

Paquerot, S. 2007. The Challenges of a Legitimate Governance of the Great Lakes and the St. Lawrence: Between Ecosystem Considerations, Diversity, and Fragmentation. *Quebec Studies*, 42(Winter), 111–31.

Picot, C., Nguyen, T.A., Roudet, A.C., and Parent-Massin, D. 2011. A Preliminary Risk Assessment of Human Exposure of Phytotoxins in Shellfish: A Review. *Human and Ecological Risk Assessment*, 17, 328–66.

Polischuk, P. 1996. *Second Life: Options and Opportunities for EPA in a Virtual World*. Washington, DC: Woodrow Wilson International Center for Scholars.

Rhodes, R. 1996. The New Governance: Governing without Government. *Political Studies Review*, 44(4), 652–57.

Rose, N. 1999. Inventiveness in Politics. *Economy and Society*, 28(3), 467–493.

Sawyer, B. 2007. *Executive Summary of Serious Games: Improving Public Policy through Game-based Learning and Simulation*. Washington, DC: Woodrow Wilson International Center for Scholars.

Sparke, M. 2002a. Between Post-colonialism and Cross-border Regionalism. *Space & Polity*, 6(2), 203–13.

Sparke, M. 2002b. Not a State, but More than a State of Mind: Cascading Cascadias and the Geoeconomics of Cross-border Regionalism. *Globalization, Regionalization and Cross-border Regions*, edited by M. Perkmann and N. Sum. New York: Palgrave Macmillan, 212–40.

Shared Waters Alliance (SWA). 2007. The Shared Waters Alliance: Proactively Addressing Water Quality in an International and Multi-Jurisdictional Watershed. Paper presented at The Georgia Basin Puget Sound International Conference, Seattle, WA.

Swyngedouw, E. 2000a. Authoritarian Governance, Power, and the Politics of Rescaling. *Environment and Planning D: Society and Space*, 18, 63–76.

Swyngedouw, E. 2000b. Elite Power, Global Forces and the Political Economy of "Global" Development. *Oxford Handbook of Economic Geography*, edited by G. Clark, M. Feldman, and M. Gertler. Oxford: University Press Oxford, 541–58.

Taylor, R. 2004. *Creating a Better World: Interpreting Global Civil Society*. Bloomfield, CT: Kumarian Press.

Tucker, Brian. 2013. Inventing the Salish Sea: Exploring the Performative Act of Place Naming off the Pacific Coast of North America. MA Thesis, Department of Geography, Victoria: University of Victoria.

Tvedt, T. 2004. Development NGOs: Actors in a Global Civil Society or in a New International Social System? *Creating a Better World: Interpreting Global Civil Society*, edited by R. Taylor. Bloomfield, CT: Kumarian Press.

Van Rooy, A. 1997. The Frontiers of Influence: NGO Lobbying at the 1974 World Food Conference, the 1992 Earth Summit and Beyond. *World Development*, 25(1), 93–114.

Worster, D. 1977. *Nature's Economy: A History of Ecological Ideas*. Cambridge: Cambridge University Press.

Chapter 5

A Basis for Bordering: Land, Migration, and Inter-Tohono O'odham Distinction Along the US–Mexico Line

Kenneth D. Madsen

Introduction

The evolution of distinction among previously cohesive border communities is an important bordering process, but one that remains largely hidden from view. Social and cultural changes are often an inadvertent by-product of greater political and economic divisions between states, but in turn they also further facilitate border-based distinctions. With a strong social and cultural basis differences along borders have the potential to become even more enduring than explicit law enforcement and border-reinforcing activities undertaken by state actors themselves. This process of social and cultural bordering is in large part effectively implemented and ultimately perpetuated by non-state actors along the border—local communities and cross-border relationships that are reluctantly and/or unwittingly drawn into the bordering process.

In this chapter I consider the case of the indigenous Tohono O'odham. The division of their traditional land base between Mexico and the United States in 1854 was not initially a barrier to interaction between the two sides, but over time national distinctions developed within Tohono O'odham society. One of the most important foundations for divergence was the Tohono O'odham relationship to land, which was detached from traditional usage and encroached upon by outsiders in two very different ways on either side of the border. In understanding how variation in Tohono O'odham relationship to the land served as a basis for internalizing the border to the point that today it serves as an us–them reference point with profound cultural and political resonance, insight is gained into the bordering process as carried out in local societies.

Many of the driving forces for bordering are often state-centric in nature, of course. Whether motivated by economic or security interests, perceptions of state-level costs and benefits dominate decisions regarding when, where, and how to control the international flow of goods and people. The fortification of sovereignty against violation or encroachment is similarly dominated by national passions. Although at the extremities of the state, borders are often "at the heart of nationalist discourse about the meaning of the nation, of arguments about who should be included in the nation and who should be excluded" (Megoran 2004: 736). They are also important institutions and symbols in perpetuating a state's territorial power in an era of globalization (Paasi 2009). The Border Patrol often takes a particularly notorious function on the front line of these processes from the perspective of US local border populations.[1]

Yet concepts of central standardized control are both challenged and perpetuated in the borderlands themselves by communities on the receiving end of state-level policies. It is in the borderlands where the nation-state as outlined by external boundaries is most tested and the alignment of nation and state most questioned. The identities of those living in borderland communities spans or ignores borders even as local residents are pushed to align more exclusively with one side or the other (Hardwick and Mansfield 2009, Martinez and Hardwick 2009, Konrad and Nicol 2011, Jones 2012). It is also in the borderlands that bordering processes designed to ensure a minimum level of national homogenization are frequently made transparent as border communities provide a basis for direct international comparison.

In the present case study I take a regional and cultural approach to the political geography of borders. By integrating the dynamics of land tenure and indigenous relations with an understanding of what comprises a geographic unit of identity and interaction, I seek to connect the physical and legal dimensions of bordering with its impact on the socio-spatial consciousness of border residents. Land and territory may provide a basis for bordering in a technical sense, but it is related cultural experiences and the internalization of distinction beyond the level of the state that accomplishes the task of social bordering. Not all who live in border regions experience the re-orientation of their land base in the way that the Tohono O'odham have, of course, but other local processes may equivalently serve as a basis for distinction and a catalyst for division. Similarly, it would be unwise to depend on a single attribute in defining an identity (Prokkola 2011: 22). Nonetheless, land has certainly played a critical role in the Tohono O'odham situation.

Semi-structured interviews with elected officials and community members undertaken as part of my dissertation research (Madsen 2005) and related research projects cumulatively provide insights on how contemporary Tohono O'odham individuals feel about the international border. Supplemental primary sources included tribal newspapers and archives at Venito Garcia Library in Sells, Arizona.[2] Several years of residence and employment on the reservation and ongoing visits and contact with individuals in the area provide a context for these issues on a more personal level.

---

1   For discussion of the Border Patrol's presence among the Tohono O'odham see Luna-Firebaugh (2005), Madsen (2005), Spears (2005), Madsen (2007), Singleton (2008), and Van Otten (2009).

2   In 2010, these files were transferred to the Tohono O'odham Cultural Center and Museum, Himdag Ki: Hekĭhu, Hemu, Im B I-Ha'ap (Culture House: Yesterday, Today, Tomorrow) in Topawa, Arizona.

Field research on this topic was informally authorized through the Tohono O'odham Nation's Executive Office in 2001 and more formally reviewed and approved by the Cultural Preservation Committee of the Tohono O'odham Legislative Council in May of 2003. I also interacted with and received approval in various forms from many of the districts (see Madsen 2005: 29–34, 223–4) that make up the reservation political system on the US side, as well as individual communities which comprise each district. It should not be inferred, however, that the results interpreted here represent an official position of the tribal government or tribal members.

Bordering as a Local Process

Whether protecting territory against intrusion by others, regulating patterns of interaction, or inspiring national allegiance, borders provide a legal justification for control. But while on a national level they may command respect as an inherent component of the contemporary nation-state (Agnew 2007), in the borderlands many residents balance borders as conceived in an absolute sense nationally against local cross-border connections. Given legal, economic, and political associations, borderland residents are influenced by their respective countries in important ways. National relationships exert constant pressure to re-direct associations that span the border and compel borderland residents to identify in more substantial ways with their domestic interiors, a process particularly noticeable among indigenous populations. Sometimes this is to the point of even writing local borderland societies out of a nation-state's history as happened with the indigenous Western Abenaki of Vermont, New Hampshire, and Quebec (Manore 2011).

A particularly elusive aspect of the alignment of nation and state is the process by which borders come to have meaning to local residents based on fundamentally different national understandings of territory and belonging. To more fully and accurately conceptualize bordering, it is incumbent to recognize the ways in which borders influence the daily lives of residents and come to be respected as authoritative among local communities. The means by which borders are internalized—not just as a legal phenomenon but as part of one's personal ethos in terms of how the world is and should be organized—is crucial to understanding the power that they possess. Nonetheless, the question remains as to where to look for evidence of bordering practices outside of state-initiated processes, how they impact particular places, and who takes part in such processes (Johnson et al. 2011: 62). Processes of distinction between "us" and the "other" are often hidden from view (Newman and Paasi 1998: 201) as part of a nebulous package of cultural bordering and recognizing some of its component parts is critical to a fuller comprehension of bordering practices.

In his biography of the Kyrgyzstan–Uzbekistan boundary, Megoran alludes to the "gradual divergence" of political and macroeconomic redirection that led to a differentiation between those two countries (2012: 472). Such factors are major driving forces in the bordering process, but differences between countries are eventually further and more forcefully manifest in the thought processes and group identities of border residents and in the mental maps with which people make sense of their surroundings. In this way culture—an agglomeration of a group's history, practices, beliefs, and material manifestations thereof—becomes a bordering process in its own right. Culture has long been used as justification for aligning nation and state, but in this chapter I refer not just to the meeting of two sets of traditions at borders as is commonly understood, but to the emergence of differences within a relatively homogenous society. Equally critical is how a society holds onto past experiences vis-à-vis international borders. By analyzing the content of a theatre play and audience reactions to its discussion about borders, Strüver (2005) argued that internal European boundaries have persisted in people's minds well past their geopolitical utility. Alternatively, Jones (2012) argues that despite an outward acceptance of the sovereign authority of India and Bangladesh, many Bengalis continue to think about national belonging in ways that transcend state boundaries half-a-century after the emergence of that particular border and despite a substantial ratcheting up of enforcement in recent years. Taking the persistence of differences and similarities between people and places as a point of departure, this chapter explores how border-based processes of distinction between states become internalized in the first place despite deeply-held notions of continuity within local communities.

For the Tohono O'odham in what is today southern Arizona and northern Sonora, as with many indigenous groups, relationship to the land is a defining component of their identity and one that often persists despite temporary absence or permanent removal. The tribe's territory is a contemporary communal expression of that relationship and is something integrally intertwined with traditional ecological knowledge, language, and even social responsibility. Territory is central to many people's national identity, of course, but for non-indigenous groups it is often in an abstract sense more removed from everyday experience and lacks association with a history which has stripped them of their greater territorial land base.

Despite significant cross-border linkages that endure to the present day (Weir and Azary 2001, Spears 2005) the Tohono O'odham relationship to the land began to be incrementally impacted by two very different concepts of native land tenure and indigenous cultural rights when traditional Tohono O'odham territory was divided between two countries with the Gadsden Purchase of 1854. As a result land has been a driving force in shifting interactions among the Tohono O'odham and re-orienting relations in two different directions. This was illustrated in a domestic context by Henderson (1991) who showed that the political relationship between American Indians and the US government was also an important component in the divergent settlement patterns of Chiricahua Apache as they were settled in distinct geographic contexts under varying social and political constraints. By contrast, the Tohono O'odham relationship to the land became differentiated *in situ*, but divided among two distinct national governments. Given an understanding of territory as a

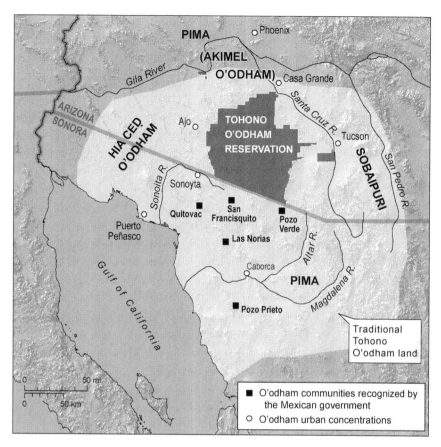

**Figure 5.1   Traditional and present-day Tohono O'odham territory
(after Erickson 1994: 17)**

Note: Cartography by B. Trapido-Lurie.

political technology (Elden 2010), varying application of that technology claimed and molded Tohono O'odham land and society on either side of the border in very particular ways.

Seasonal Land Use and its Legacy

The manner in which O'odham-speaking people have been named and categorized and their territory constrained in scope by successive Spanish, Mexican, Anglo, and even Apache regimes provides a brief introduction to the dynamics of indigenous regional identity in the Sonoran Desert. Translated into English the term "Tohono O'odham" refers to "Desert People." This group was historically known throughout most of its post-Contact history as the Papago, a term that can be traced to initial contact with Spaniards. The availability of water defined seasonal migration of the Tohono O'odham and this was a dynamic not easily constrained in an arid climate with the arrival outside groups accustomed to more permanent settlement patterns. In the winter, "well" (natural spring) villages were inhabited at the base of mountains. Upon arrival of the summer monsoon rains that provided moisture for crops and residents, settlements would relocate to basin "field" villages (Underhill 1939: 57, Lopez, Reader, and Buseck 2002). A village's traditional role in this seasonal arrangement is easily discerned based on its placement within the basin and range topography of the Sonoran Desert and institutionalized in some of the names by which many places are known yet today: Gu Oidak (Big Field(s)), Newfield, Ak Chin (Arroyo Mouth, where summer rains can be put to use for agriculture), Pozo Verde (S-cedagĭ Wahia, or Green Well), Sif Oidak (Bitter Field), Covered Wells, Vaya Chin (Well Mouth), and even Tucson (S-cuk Şon, Black Water Spring).[3] Substantial branching might occur as a group broke up into multiple villages only to reunite several months later or as settlements grew larger than a single village could support (Hoover 1935, Jones 1969: 171).

The Pima (also known as "Akimel O'odham" or "River People") were differentiated through their association with more permanent settlements along rivers, while the Hia-Ceḍ O'odham ("Sand Papago") to the west identified with more nomadic movement across starker desert landscapes. The later setting aside of reservation lands and regulation of membership and identity in the US reinforced these divisions. Also O'odham-speaking, but later to lose a unique identity, the Spanish recognized Sobaipuris to the east along the San Pedro and Santa Cruz river valleys. Although differences between the Pima and the Papago were probably no greater than internal variation within each group, such categories became engrained in the era of European colonization. In the reduction of traditional lands to their present extent on the US side, Papago/Tohono O'odham took over as the dominant identity of south central Arizona and northern Sonora even in areas where Pima and Sobaipuri had been dominant. The Hia-Ceḍ O'odham, retaining a unique identity, were consigned to the Papago Reservation at the time of its creation.[4]

------

3   For a brief listing of O'odham place names see Saxton, Saxton, and Enos (1983: 128–38). A more complete but older listing is available in an internal report by the Papago Indian Agency (c. 1942). Confusing the issue for many outsiders is the use of multiple names for single villages using different languages, translations, orthographies, and dialects. Some locations may also have traditional, modern, or Saint names by which they are known. On occasion a "well" name may refer to a government-dug variety rather than a natural spring.

4   Fontana and Owens (2004) provide a brief introduction to the diversity and consolidation of various O'odham groups. For an excellent and in-depth discussion of the consolidation of Tohonon O'odham identity, see Booth (2000). The official tribal history (Erickson 1994) is also a good source. Not previously acknowledged within the formal

Their seasonal pattern of residence provided the Tohono O'odham with identity and sustenance. Field villages allowed the cultivation of tepary beans, corn, squash, melons, chilies, and sorghum to supplement hunting and gathering, whereas the latter dominated during time spent in the well villages (Underhill 1939: 57, Nabhan 1985, Erickson 1994: 8–10, Lopez, Reader, and Buseck 2002, Buseck 2003, Fazzino 2008). The harvesting of fruit from saguaros, probably the single most important food to Tohono O'odham identity, took place in early summer to a certain degree in both locations. Occasional visits were also made to more permanent Akimel O'odham settlements along major rivers to exchange goods and labor, especially in times of drought (Erickson 1994: 17).

Seasonal interactions between villages were disrupted during the Apache threat in the late 1800s when many villages on the eastern edge of traditional O'odham territory were left empty for extended periods of time as the O'odham contracted into fewer villages for defensive purposes. After this time, however, seasonal "village systems" once again began to disperse (Hoover 1935: 262, Hackenberg 1964: 291–9). Traditional seasonal migration patterns and accompanying village system expansion began to wane once again in the 1900s (Wilson 1972: 115, 131) as locations with access to permanent water and additional utility connections became accepted and even favored. By the time the Papago Tribe was recognized as a domestic political entity with a defined land base in 1937, for several decades the federal government had been digging permanent wells and establishing and improving *charcos* (ponds) to retain rain water (Erickson 1994: 103, 135, Lewis 1994: 149, Booth 2000: 118). These efforts altered the need for migration and changed settlement patterns to give rise to new village place names like Wainomĭ Ke:k (Iron Pipe Standing) and Charco 27.[5] Roads and other infrastructure are less determinative in Tohono O'odham settlement patterns than in many places in the US but they are still important. The lack of modern infrastructure in some locations resulted in a gradual or even complete loss of population in "permanent" or "primary" residents. While these two words may seem straightforward to outsiders, they are loaded with misleading cultural assumptions not necessarily shared by the O'odham.

Although residences were not fixed in the same sense that they are in many other societies (Hoover 1935, Oblasser 1936, Hackenberg 1964) for the Tohono O'odham a place's intrinsic importance as home was and is maintained despite absence. This was true not only between seasons, but over years and even

---

political structure of the Tohono O'odham Nation, a new Hia-Ced District was approved by the tribe in October 2012.

5   Although in general agreement with the field-well dichotomy, Jones (1969) alternatively identifies many locations as *charcos* for the period he studied, 1900–1920. *Charcos* certainly pre-dated this era, but increased in importance at this time. Place names utilizing the O'odham terms "Wo'o" or "Vo" indicate an attachment to this type of water source.

generations. Current and formerly-occupied villages remain bound to each other through history and culture. To others cycles of absence have the appearance of permanence but they were never an indication of abandonment for the Tohono O'odham themselves. Nonetheless, Mexicans and Americans often viewed it as such (Jones 1969: 103, Nabhan 1982: 69–71, Erickson 1994: 77) and that perception was accentuated as seasonal migration gave way to longer periods of absence. Even today a location does not have to be lived in on a permanent basis to maintain the significance of home, and traditional experiences with seasonal residences continue to color contemporary Tohono O'odham settlement and commuting patterns, a pattern also observed by Henderson (1991: 12–14) with regard to the history of the Chiricahua Apache.

Today, with long distances to travel to the center of political, educational, and economic activity in the reservation capital of Sells, a number of people maintain two places of residence—one in Sells during the week and one at "home" for weekends, family events, and cultivation of community and identity. Other tribal members prefer to live in Tucson from where they commute into Sells daily, even if they have a house on the reservation that remains vacant or is watched over by someone else. Although a shortage of viable housing on the reservation exists, having two homes is not considered unusual and when opportunities arise for new housing or government infrastructure to be built community pressure exists to have it dispersed across the reservation rather than centralized for easy access. This dynamic has not translated into a renaissance of village migration in the seasonal sense, but it does reflect a tradition of preferred settlement dispersal based on community identity rather than concentration for convenience and often results in daily, weekly, or career-cycle movements that straddle modern and traditional demands for living arrangements.

This pattern was recognized a half-century ago when Kelly wrote "Many of the villagers in San Miguel, Vamori and the settlement of Tecolote have temporary homes in these villages and thus spend most of their time living in Sells where they work for wages. Oftentimes the 'temporary' residence in Sells is in fact the 'permanent' one" (1963: 62). Nonetheless, Kelly imposed an outsider's understanding of the situation when he wrote of Comobabi that there were "fifteen houses here in good condition and at least six to eight of these are inhabited seasonally. This village is best described as seasonally 'vacant'" (1963: 66). If forced to put it in terms comprehensible to an outsider, a Tohono O'odham individual would probably describe the figurative glass as half-full rather than half-empty! One's home village is vastly more important than where one lives (spends most of their time) when introductions are made to others and, even if someone has *never* lived there, until recently it was one's only option for voting in tribal elections. Tohono O'odham personal and political identity is rooted in family and history rather than contemporary geographic proximity. Among my own contacts on the Tohono O'odham reservation, I have known some individuals for years before realizing that for the most part they lived somewhere other than their declared home.

The traditional Tohono O'odham dual-residence arrangement clashed with patterns of US and Mexican land use. How land was used or not used by the Tohono O'odham was a critical issue in establishment of the main ("Sells") body of the reservation. As debate flared after the initial declaration establishing this section by presidential proclamation in 1916, the Arizona Legislature questioned the move to create such a vast reservation. The state land commissioner reported that "altogether there were only about 4,500 of the Indians, many if not most of them resided in Sonora, moving back and forth across the line. They could not utilize the land—it could not be said that they could occupy it" (*The Arizona Republican* 1917). Ironically, a new type of seasonal land use led to one of the Tohono O'odham's largest allies in the fight for substantial reservation lands. The cotton industry lobbied forcefully for a reservation that would provide a base for its many Papago employees during the off-season (Erickson 1994: 110, Booth 2000: 124–7, Meeks 2007: 56–7). Reservation advocates went to great lengths to justify rights to the land by demonstrating that Papagos at the time "were not nomadic but possessed fixed homes" (Booth 2000: 137), an argument necessitated by outsiders' belief that the two village Papago somehow did not deserve land because of their seasonally itinerant lifestyle. Partly as a result of debates over permanent occupancy, for many years the Papago were granted a reservation much smaller than originally proposed (Erickson 1994: 107, Booth 2000: 142–55).

Eventually active seasonal migration in its various incarnations faded and Tohono O'odham families settled into the villages with which they were familiar. Absence has yet to become synonymous with abandonment in O'odham culture, however, and close personal and territorial relationships between locations historically related as part of the same village system continue to exist. While some young people today may not immediately recognize the personal connections as remnants of an earlier seasonal migration pattern, they make up a significant element of *'i:migĭ*, traditional family and social relationships that are critical to O'odham identity. Although seasonal migration spanned the border and acted to blur its sharpness historically, these relationships have been overshadowed in recent years by a focus on enforcement of the official boundary itself in what I have elsewhere described as an alignment of borders as functionally interpreted on the local scale (Madsen 2014).

Seasonal and other forms of circular migration historically served to attach Tohono O'odham individuals and communities to each other as well as distinguish them from other O'odham groups. In terms of interactions with non-O'odham, however, the legacy of these patterns ultimately had very different impacts on either side of the border. Most significantly and to the detriment of Tohono O'odham occupancy, seasonal usage provided openings for outside encroachment as land was perceived as abandoned. Seasonal land use dynamics were controversial when allocating land use for exclusive Tohono O'odham usage in the United States, but encroachment was largely halted with the formal creation of the reservation on the US side. On the Mexican side the perception of land abandonment lingers today and Mexicans continue to settle on and utilize traditional O'odham territory—a perception aggravated by out-migration.

Out-migration, Residential Tenure, and the Cultural Importance of Land

Like many Native Americans, the Tohono O'odham have an affinity for the land that borders on spirituality and does not lend itself well to a brief academic analysis. It is O'odham *ha-jewe̗dga*, land that belongs to the O'odham and where the Creator I'itoi roams. The importance of land goes beyond just a place for O'odham to live or earn a living; it harbors cultural heritage sites and common O'odham legends are set in both Arizona and Sonora. Stories told in the wintertime share lessons based on the geography of the region. As Basso (1996) observed with the Western Apache, wisdom does indeed reside in these places. Even if an O'odham individual has never been to these locations, he or she knows of their importance. Communities are comforted knowing that the land is still in O'odham hands. Today permanent legal rights are ideal, but informal possession is also useful. Heritage locations can remain central places to the O'odham even as they appear to outsiders as underutilized land. As a consequence, protections over such places are weakened as O'odham migrate out. In one interview a person with an O'odham grandparent from Imuris, Sonora—a town that no longer has an identifiable O'odham presence—lightheartedly reminded me that O'odham place connections are not just in passing.

> [W]e went to Mexico with my wife, her niece and her nephew, and my nephew. And I was telling them about it and when we got to cross the border, going into Mexico and we stopped at Imuris and I told them … 'we're home, we're home.' And they were all laughing because they understood at the time that I knew there was a record of their family coming from Imuris.

Accustomed to seasonal migration, O'odham movement across the border was historically not an uncomfortable leap for many. Seasonal or permanent migration was often undertaken to whichever side provided the best living conditions (Nolasco A. 1965: 447). Although Papagos had been drawn to Spanish missions or Mexican settlements as laborers for centuries and in the late 1800s some O'odham went south into Mexico to work in the fields and get away from the arrival of Anglo settlers (Erickson 1994, Fontana n.d.: 191–2), dominant flows eventually turned north and much of it became permanent. Several have already written about how many O'odham migrated across the border to Arizona over time as Papagos were crowded out of Sonora and job opportunities became more plentiful in the United States (Joseph, Spicer, and Chesky 1949: 22, Spicer 1962: 133, 138, Gomezcésar Hernández 1997: 52–3, Booth 2000: 87–9). Whereas O'odham from Mexico went to work in a variety of locations north of the reservation, the flow to the area east of Casa Grande to work as agricultural laborers was particularly strong (Waddell 1969). Many O'odham from the US side were also departing the reservation to look for work (Meeks 2003), suggesting that on-reservation areas were not viable destinations in the mid-1900s for O'odham coming from Mexico. O'odham from Mexico migrating to non-reservation destinations in the US did not generally establish substantial and lasting geographic connections with the reservation base on the US side.

Relationship to the Land as a Bordering Process

Among O'odham populations of Mexican origin three distinct groups developed based on their heritage, geography, and perceived levels of acculturation. Those who maintained connections across the international border within shared village systems were the most integrated with their US counterparts. Less connected were those who undertook migration to off-reservation locations in the United States or to predominantly non-O'odham urban areas in Mexico. Among tribal members today the acronym "OIM" as shorthand for O'odham in Mexico is used to refer to all three groups and their descendants, although some OIM who end up on the US reservation may shed that affiliation by establishing themselves as long-term members of one of the reservation districts.

These distinctions have repercussions today as both O'odham residing in or with origins in Mexico comprise a large segment of the membership which is designated as having a "no district" land base.[6] This status came with a certain degree of political disenfranchisement within the tribe in the past given that the only way to vote was through one's district of membership (*Runner* 2003) but this situation appears to be changing. Although application to the OIM situation is not yet clear, a new election ordinance passed in August 2013 now allows members to register to vote outside of their district of membership (Tohono O'odham Legislative Branch 2013). Nonetheless, there is still no direct representation of the tribe's Mexican membership as an identifiable unit in the Tohono O'odham Legislative Council.

In contrast, US O'odham who migrate off reservation are likely to maintain an affiliation with their district of origin. This bifurcation has its origins in the lack of a geographic base for O'odham originating in Mexico within the constitutionally-based tribal government and has contributed to a deeper us–them dichotomy between O'odham in Mexico and US Tohono O'odham members. Although they may live in what was traditional Tohono O'odham territory, O'odham in Mexico (whether physically in Mexico or living in the US) are a people without a formal land base in the contemporary Tohono O'odham political sense. Given the cultural

---

6 According to a 2001 tribal publication, approximately 1,400 tribal members resided in Mexico at that time (Castillo and Cowan 2001: 10). In 2007 just over 2,000 or 8 percent of the tribal enrollment of 28,000 members was designated as "no district" (Tohono O'odham Nation 2013). While there is substantial overlap between these two categorizations, however, they are not equivalent.

importance of land in Tohono O'odham society this serves as the subtle but significant basis for distinction between the two sides.

Relationship to the Land in Sonora

Historically O'odham land immediately south of the border remained relatively unsettled by non-O'odham Mexicans due in part to the O'odham presence there. Also serving as a deterrent was the area's relative isolation in terms of roads, proximity to other settlements, and in some places water. Although roads remain poor and infrastructure such as modern wells and utilities is often lacking even today, demand for land by Mexican ranchers has increased. Smugglers also see the area as desirable for illicit staging operations with certain infrastructural shortcomings and jurisdictional isolation from both the US and the rest of Mexico recast as advantages. Land not actively in use is a relatively easy target and squatters are able to file for title after a few years of occupation. As a result, informal and communal usage of the land by O'odham living in Mexico has been usurped by comparatively more aggressive and individually-oriented Mexicans. Confrontation, furthermore, is not a traditional O'odham approach to conflict and interlopers are sometimes left to themselves while they reside in an O'odham village to establish their claim.

A handful of O'odham villages and their land in Mexico were organized communally under land reform programs over the last century as Papago *ejidos* and *comunidades*. This compensates somewhat for the lack of a larger political land base but these are relatively small areas and the protection offered is less comprehensive than the reservation's status in Arizona. Furthermore, neoliberal reforms in 1992 lessened their protection by allowing for privatization of *ejidos*, even though in the past such lands technically provided only usufruct rights and were not expected by the Mexican government to sit idle for cultural and historical value anyway. Rather, they were intended to be occupied and actively productive economically. In this way, Mexican land tenure laws conflict with the O'odham philosophy of land being valuable in its own right, without having to serve as a permanent residence or be economically productive (Barnett 1989: 17).

Although the *mestizo* bloodline has traditionally been idealized, ideas about autonomy and a protected land base for indigenous people in Mexico are relatively new given that almost all Mexicans believe themselves to be descended from an indigenous past. If the average citizen were to ponder the topic of ethnic identity, he or she might wonder why anyone deserves to be recognized as more indigenous than the rest of the country. Everyone, after all, is part *indio* by virtue of being *mestizo*, and many so-called indigenous people have a mixture of non-native blood anyway. In the US, by contrast, for all its bureaucratic faults and economic challenges, the reservation system has fostered a tight sense of community for a relatively sovereign Tohono O'odham government and a secure land base regardless of use or economic productivity. In the US the O'odham also have a unique government-to-government relationship with increasingly strong legal standing. Many US O'odham have come to take such protection for granted on both sides of the border and this has led to frustration over Mexican land issues for some and a misleading sense of security for others.

Though many Mexican O'odham have a basic understanding of their government's system of land ownership, few on the United States side of the line—and sometimes even those with heritage in Mexican border villages—are aware of the related details (Barnett 1989: 8). Despite a desire to help protect Mexican O'odham lands from further encroachment, incomplete understandings of the political and legal environment in Mexico hampers effective cooperation between the two sides on this issue. Furthermore, lacking the political organization and clout of the tribal government which is based on the US side, O'odham possession of land in Mexico has been diminished and remaining locations are at risk. Some transplants now residing elsewhere remain in touch with their rural Mexican origins, but the reservation or the city increasingly defines their existence and land left behind is vulnerable to outright appropriation by others. In the context of O'odham geography, families and individuals may remain within their larger village kinship grouping (interpreted and adapted to include nearby urban areas) but the result has been a void of ongoing residents in many rural O'odham villages that lie within Mexican territory. While these places continue to be important ceremonial sites and part-time residences, fewer and fewer O'odham live in them on a regular basis.

Relationship to the Land in Arizona

Reservation status means that rural-to-urban migration on the Arizona side is experienced very differently. Together with local economic pressures to urbanize and off-reservation economic opportunities, official relocation programs were sponsored by the US federal government starting in the 1950s. These forces were tempered by the tax-free status of reservation land, strong cultural ties to the land, the communal orientation to O'odham society, and the importance of *'i:migĭ*, but nonetheless today the number of off-reservation individuals rivals that of the on-reservation population.[7] As in Mexico, what it means to live away from one's traditional land base and still be Native is a point of contention, but socially and morally the center of Tohono O'odham culture and government in the United States remains rural and reservation-based. Some tribal members have gone to the city only temporarily for education and work, others commute back on a regular basis. Even among those who move farther afield, reservation lands await their

---

7   In 2007, only 48 percent of the tribe's membership resided on US reservation lands (Tohono O'odham Nation 2013). Counting only US members from the calculation, by removing Mexican-based members (as enumerated in Castillo and Cowan 2001: 10), raises that to 50 percent.

return should they make the decision to do so. Much reduced from pre-Contact times, US land that remains in O'odham hands today is watched over culturally by members who remain and legally by virtue of the area's trust status with the federal government. It remains solidly protected from further encroachment by non-O'odham settlers.

If it is true that places can hold different meanings for diverse groups simultaneously (Toupal 2001, Ferguson and Colwell-Chanthaphonh 2006), possession by others does not inherently break the bond which links the O'odham to significant places in southern Arizona and northwestern Sonora. It is, however, a serious challenge. Landscapes in southern Arizona which are not under direct O'odham control today are still O'odham places to a certain degree, made acceptable by the preservation of a large amount of land and a number of important sites within the reservation. In addition, US public land administrators are obligated to consider Native perspectives through specific formal channels when developing land use plans and under many circumstances even private developers are duty-bound to watch out for and take care of archaeological sites they encounter. There is even a legal obligation to repatriate uncovered human remains to the tribe deemed to be their closest living relatives. Recourse by the Tohono O'odham Nation is available through the courts and can be done in a familiar legal environment. Culturally and legally, historical villages such as Arivaca and Tumacácori (see Wilbur-Cruce 1987, Sheridan 2006), long since taken over by non-O'odham, remain more accessible than do lands in Sonora.

Relationship to the Land in National Context

Tohono O'odham seasonal land use, urbanization, and other forms of connecting and remaining connected to the land have been situated in two very different national contexts over the last century and a half. Political, economic, and cultural systems manipulated and constrained the Tohono O'odham and their relationship to the land in two very different ways. From the US O'odham perspective, there is even an undertone of being more responsible in this respect whereas their Mexican brethren have failed to adequately protect their land. When understood this way, land and migration—two central concepts of traditional Tohono O'odham identity—have served to differentiate the two sides and isolate them from each other rather than serving as common experiences.

By comparison to the US, protections over historic and cultural sites are not as extensive in Mexico, there is no tribal-wide organization based there with political influence comparable to the US-based Tohono O'odham Nation, and the unfamiliar international legal environment hampers efforts by the US side to fight for such protection. Remaining O'odham sites in Sonora are more critical than one might imagine because their loss would mean the outward disappearance of a traditional presence in that state. Efforts have been made to preserve access to land in Mexico, and feelings of stewardship over[8] Mexican O'odham lands remain strong on both sides of the border. In the end, however, legal protections are not as vibrant or enduring as on the US side. Together with Mexican members' isolation from the tribe's greater contemporary territorial base and its US-based governing structure, these differential legal relationships to the land figure prominently in distinguishing between the two sides.

Cultural Distinction and Furtherance of the Bordering Process

Given local seasonal migration patterns certain remnant villages along the border in Mexico remained largely integrated with and indirectly sustained by contact with US-based O'odham while at the same time gradually becoming isolated from other O'odham communities in Mexico. Internal village system migration patterns served as an alternate and more organic basis for international distinctions (Madsen 2014). In general *border* OIM followed the acculturation patterns of the rest of their US-based village systems, even as *non-border* OIM became differentiated by US-based O'odham as more acculturated to non-O'odham Mexican society.[9] From a more neutral perspective, each group has been heavily influenced to varying degrees by non-O'odham society and might simply be described as different. Such broad generalizations certainly do not hold up in all situations, but they are nonetheless useful indicators in understanding the dominant orientation of each side. Furthermore, some individuals may transcend this division and the categories are fluid depending on the context considered.

Intermarriage and the lack of a solid land base in Mexico were important contributing factors to *mestizaje*, the process of becoming biologically and culturally integrated into broader Mexican society. Some individuals in northern Sonora know they have Papago forebears but feel no cultural affiliation to the tribe and have never considered that they might be able to apply for membership. Not considered Papago by themselves or anyone else, they are today simply Mexican with a slightly identifiable Indian heritage. Others have lost track of that

---

8   The Western terminology of "ownership" is also used sometimes although formal title may not always be held. In return, from an indigenous perspective, many O'odham also consider themselves as belonging to the land.

9   OIM who have migrated to off-reservation locations in Arizona fall somewhere between these two (depending on specific circumstances) and are not within the scope of this chapter. It should be noted that I use acculturation to refer to an increase in ties between the O'odham and the dominant culture, economy, and political life of each country. Similar to Nagel's (2009) understanding of assimilation of immigrant groups, my emphasis is on the process of making sameness—the establishment of a common bond or experience—rather than an end product of "Americans" and "Mexicans" who no longer see themselves as O'odham.

element of their past completely. Although the *mestizo* is valorized in Mexican history, there has also been a tendency to look down on *indios* proper, and it was convenient for many generations to simply let go of that part of one's history. Today towns and cities with O'odham-inspired place names like Tubutama (probably *Jewed Dam* or earth above, mountaintop), Oquito (perhaps derived from *Giho Do'ag* or Burden Basket Mountain), Bamuri (*Vamori* or swamp), and Caborca (*Ga:wolk* or hill) to the east and south of today's Tohono O'odham core are already overwhelmingly *mestizo* in identity.

Even among communities in Sonora that do retain an explicit O'odham identity, there are acculturation challenges. Today preservation of the O'odham language is one of the most critical identity issues being fought on the US side, but this battle is commonly perceived by those living on the reservation to have already been lost in Mexico where language frequently serves as a defining factor of who is *indigena* and who is *mestizo*. Among members living on the reservation itself, a decline in O'odham language use is also clearly evident but still several generations removed from approaching the situation in Sonora (Lopez 2004, Madsen 2005: 81, Naranjo 2011: 17).

The English–Spanish divide is also problematic and contributes to an inability to communicate, further increasing the social distance US O'odham feel with those across the border. Even if a Sonoran member meets the political and legal (i.e. blood quantum) criteria for official tribal enrollment, cultural belonging can be brought into question with the lack of a common language. Individuals who have lived in Phoenix for generations might not speak O'odham either, but they do speak a common language (English) and can be understood—they are not as "foreign." The same could be said for someone who lives in Sells but does not speak O'odham. These individuals have the alternate qualification of being physically present on land which remains in O'odham hands and participating in everyday local activities which, by their geographic nature of happening on the reservation, are O'odham activities.

When Mexican members speak Spanish at the local Indian Health Services, tribal offices, or the Basha's grocery store in Sells they stand out as different. Even OIM mannerisms are unfamiliar and their attitudes and actions come across as distinctively assertive. These individuals do not see themselves as any less O'odham because they are Mexican, just as those living on the US side do not consider themselves any less O'odham because they are American. As the ones who are visitors to the Nation's center-of-operations, however, Mexican O'odham are more likely to be evaluated in a critical light, especially when a familial connection is not clearly present. In Martínez's (1994) border typology, the Mexican O'odham visiting Sells are binational consumers, perhaps unaware that most US O'odham they meet are exhibiting uniculturalist or even nationalist borderlander characteristics in response to the encounter.

Beyond interactions with their respective fellow citizens and outward behavior, deeper differences are also manifest in social recognition of the other as the same kind of O'odham. Whereas most O'odham in the US look to the Baboquivari

Mountains as the home of I'itoi, in Mexico the Black Mountains north of Puerto Peñasco fill this role. A few people are familiar with these varying understandings and the existence of I'itoi's "second home" (depending on one's perspective) across the border but do not give it much thought unless queried.

The border became a convenient marker for distinction in other ways as well. In one version of a traditional story as published by Saxton and Saxton, the legend of *Ho'ok Oks* ("Monster Lady") is closed with the phrase "We Tohono O'odham end the story here, but the Sonoran O'odham go on" (1973: 261).[10] Although difficult to generalize from such a short statement, this is a subtle indication that indeed there is a process of internal distinction based on the presence of the international border. There are also situations where last names have been changed as O'odham families moved from Sonoran border villages to the United States. Some of this may be attributed to the utilization of two family names according to Mexican tradition and the need to fit this into a US model, but this does not explain all situations. Such name adjustments further reflect the process of differentiation that some underwent as they crossed the border—even as they stayed within the Tohono O'odham context—and an example of how that border reinforced a perception of difference between the two sides.

Among the Tohono O'odham "The Gate" is a well-known traditional border crossing for traveling from one country to the other and where the two sides meet and interact. Similar to border towns elsewhere fulfilling an entertainment role and experience with the "other," "The Gate" serves as a bridge between cultures while simultaneously reinforcing difference. Although certainly less contrived than border tourist districts described by Arreola and Curtis (1993: 90–96), peddlers and food vendors from Mexico sell their goods to US O'odham without their guests ever losing sight of the borderline. As a toponym "The Gate" reflects a link between the two sides, but in the Tohono O'odham language the term used—*Kolhai Ki:jg*—refers more specifically to an opening in the fence. The defining factor is the existence of the broader fence along the border, initially constructed for keeping cattle in place (Erickson 1994: 116) and more recently rebuilt to serve as a vehicle security barrier. Although targeted at outsiders, this fence now serves as a manifestation of outside political control over traditional Tohono O'odham land and the social distinctions that have sprung from this reality. Access for O'odham is generally open but even this is subject to changing interpretations of native sovereignty and central control as if to emphasize distinctions between the two sides, the marginal nature of cross-border community, and the ultimate authority of the state.

---

10   Translation by the author with assistance from O'odham friends and colleagues. The original phrase on which it was based is "Ahchim Tohono O'othham ach ia kuhugith. K washaba g So-nohla t am o'othham ba'ich ha'ichu ep a'aga." This is translated by Saxton and Saxton as "We Desert People end the story here, but the people on the Mexican side of the border go on."

Participation in cultural and religious activities is another internal distinguishing characteristic. Reservation residents are well attuned to the geographic subtleties of acculturation. Cases which illustrate differential attachments to the land and interaction with non-O'odham society include living on the north side where the children go to school off the reservation in Casa Grande, working in Sells where the contemporary sometimes overshadows the traditional and non-O'odham have a greater presence, living in the relatively conservative western districts, or participating in certain ceremonies that are maintained in Big Fields or Santa Rosa. These are all examples of various levels of acculturation, non-acculturation, and even resistance to the outside world within those who live on the reservation. Above all, however, non-border OIM are often seen as having lost the most O'odham cultural identifiers. This may be unfair to a group which exists on the margins of reservation-based Tohono O'odham society, but nonetheless such perceptions reflect parallel lives and do little to preserve or build up a common existence.

There are strong feelings by some on the US side that non-border OIM are simply too far removed from contemporary definitions of what it means to be O'odham to be effectively incorporated under a common identity. The comments of one O'odham individual living on the US side summarizes this feeling in respect to OIM who are seeking to become official members of the US-recognized tribal structure:

> Yes, they are just different. I mean, they are human beings too. I realize that, but as far as cultures being that they mainly grew up as Mexican and now they are trying to be members of this Nation, how are they going to know, or how are they going to learn, you know, this is a lot of questions.

Quitovac stands out as a Mexican village that retains a substantial O'odham identity, a status sustained in part due to its continued occupancy, the presence of a school catering to indigenous students, and views of US-resident tribal members toward this place and its ceremonial importance. Proximity to a major road would seem to be a challenge in terms of maintaining a unique indigenous identity within broader Mexican society, but also provides a regional and international connection to other O'odham. Even so, cultural heritage and legitimacy are provided as much by place as population. Members living in the western part of the reservation in the US play an important role in the leadership of cultural events in Quitovac (Weir and Azary 2001), a phenomenon in keeping with US-based Tohono O'odham generally assuming the tribe's cultural mantle. Even more significantly this particular village serves as a central gathering place for a larger village system wherein migration was not historically facilitated through easy connections to the US side. In this way Quitovac serves a connecting role among OIM who continue to identify as Papago/Tohono O'odham.

OIM also have their concerns about those on the US side. Almost half-a-century ago, a Mexican anthropologist noted that Sonoran Papagos were concerned about their US counterparts being pressed into national military service and suffering *"las restricciones de vivir en una reservación, lo cual no es de su agrado"* (Nolasco A. 1965: 447).[11] Understanding how the US tribal government operates and relates to them as Mexican members, communicating their concerns effectively across national contexts, and even understanding such issues as why the reservation prohibits alcohol sales are puzzling. Based on years of interaction focused in different directions, today O'odham living in Arizona and Sonora have fundamentally different ways of looking at the issues with which each side is faced.

The Festival of St Francis in early October is one of the more significant cultural events in the O'odham calendar. This festival's origins and strength originate from and remain based out of Magdalena, Sonora, and it is a major cultural event common to O'odham on both sides of the border that involves travel and pilgrimage into Mexico (Fontana 1981: 82–4, Nabhan 1982: 111–19, Griffith 1992: esp. ch. 3). Even so, in some ways US O'odham here are like Hispanic attendees from nearby Tucson: connected to the specific site by various intensities of heritage, geography, history, and spirituality, but culturally they remain visitors to the country. As early as 1947, US O'odham traveling beyond the border on pilgrimage to Magdalena would use their trip as "a time for recalling when the Sonora lands belonged to their ancestors" (*Arizona Daily Star* 3 October 1947 as cited in Schulze 2008: 202). Today when celebrating in Mexico the two groups often interact in separate spheres and it is less common for OIM to reciprocate by participating in celebrations on the Arizona side. Language is a contributing factor, but travel restrictions and the perception of each other as outsiders are also important issues. Even in a setting where one might expect to find remnants of cross-border unity, division and distinction based upon different experiences on either side of the US–Mexico border has crept into and now define intra-O'odham relationships.

Conclusion

With the creation of a reservation in the US after half-a-century of control and non-Native encroachment, Tohono O'odham identity in Arizona coalesced around a secure, territorially fixed land base. For all the shortcomings of the reservation system, and indeed there are many, this system of indigenous land tenure eventually came to serve as a mechanism to buffer Tohono O'odham culture and identity from excessive outside cultural and economic influences. By contrast, Mexican Tohono O'odham land was broken up into small isolated communities and as a result their identity as indigenous people became less centralized and therefore less easily recognizable by others, including Tohono O'odham on the US side. Lacking an equivalent means of protecting indigenous lands, expansion of Mexican settlements continues to encroach on O'odham residences in Sonora

---

11 "the restrictions of living on a reservation, which is not to their liking."

today. Land is not the only reason Tohono O'odham on either side of the border grew apart, but it is a major factor in their respective re-orientation.

Traditionally, Tohono O'odham identity was more based on local communities than pan-Tohono O'odham categorizations. As tribal identities those categories were solidified post-Contact, so it is not a surprise that Tohono O'odham sub-groups evolved independently based on the international border and a diverging relationship the land. As an indigenous population the Tohono O'odham did not come to the table with a proclivity to favor one side or the other. Instead, national systems of land ownership and forces of political integration and isolation influenced each side independently. As changes became manifest in Tohono O'odham society, however, culture became a bordering process in its own right and exacerbated the differences the two sides saw in each other. At a certain level O'odham in the US and O'odham in Mexico both maintain a unique indigenous identity reflecting the persistence of native presence within the nation-state system, but a closer inspection reveals that in general these two groups have been increasingly re-oriented toward their respective external national identities to the point that some question the indigeneity of fellow O'odham across the line. Even though both sides remain uniquely O'odham within their own countries, divergent experiences as a component part of two very different nation-states has exacerbated the otherness the two sides see in each other.

While the case of the Tohono O'odham is unique, it provides insight to the bordering process of differentiating between distinctive national identities. The situation also helps us understand colonially-imposed divisions around the world where the status quo of borders is staunchly defended. Finally it highlights the role of non-state entities (in this case local social norms and experiences) in the bordering process. Even where historically imposed borders egregiously cut across culturally unified groups, significant differences often subsequently develop beyond those undertaken directly by the state. These differences may be overlooked by outsiders and even remain un-prioritized by some border groups or individuals themselves, but over time can serve as a basis for reifying distinctions once prioritized only by state actors.

References

Agnew, J. 2007. No Borders, No Nations: Making Greece in Macedonia. *Annals of the Association of American Geographers*, 97(2), 398–422.

Arizona Republican, The. 1917. More Space Needed at Capitol; Legislature Talks Appropriation. 13 January: 1.

Arreola, D.D. and Curtis, J.R. 1993. *The Mexican Border Cities: Landscape Anatomy and Place Personality*. Tucson: University of Arizona Press.

Barnett, G.S. 1989. Report Regarding the Tohono and Hia-Ced O'odham of Mexico Indigenous Peoples' Loss of Their Land, Violations of Convention 107 of the ILO, Violations of Treaty Rights, and The Lack of Protection for Cultural and

Religious Rights of The O'odham of Mexico and the United States. Tucson, Arizona. 9 October.

Basso, K.H. 1996. *Wisdom Sits in Places: Landscape and Language Among the Western Apache.* Albuquerque: University of New Mexico Press.

Booth, P.M. 2000. Creation of a Nation: The Development of the Tohono O'odham Political Culture, 1900–1937. PhD Diss., History. West Lafayette: Purdue University.

Buseck, P.M. 2003. Tohono O'odham Agriculture and Traditional Foods: Revitalizing a Community Food System to Help Prevent and Treat Diabetes. Thesis, International Agricultural Development. Davis, CA: University of California.

Castillo, G. and Cowan, M. (eds). 2001. *It's Not Our Fault—The Case for Amending Present Nationality Law to Make All Members of the Tohono O'odham Nation United States Citizens, Now and Forever.* Sells, Arizona: Tohono O'odham Nation, Executive Branch.

Elden, S. 2010. Land, Terrain, Territory. *Progress in Human Geography,* 34(6), 799–817.

Erickson, W.P. 1994. *Sharing the Desert: the Tohono O'odham in History.* Tucson: University of Arizona Press.

Fazzino II, D.V. 2008. Traditional Food Security: Tohono O'odham Traditional Foods in Transition. PhD Diss., Anthropology. Gainesville: University of Florida.

Ferguson, T.J. and Colwell-Chanthaphonh, C. 2006. *History is in the Land: Multivocal Tribal Traditions in Arizona's San Pedro Valley.* Tucson: University of Arizona Press.

Fontana, B.L. n.d. The Papago Tribe of Arizona v. United States of America. As reprinted in *Papago Indians III* (1974), edited by D.A. Horr. New York: Garland Publishing Inc., 151–226.

Fontana, B.L. 1981. *Of Earth and Little Rain: The Papago Indians.* Flagstaff: Northland Press.

Fontana, B.L. and Owens, M.U. 2004. An Annotated Bibliography of the Tohono O'odham (Papago Indians). [Online: Tumacácori National Historic Park, May 2004]. Available at: http://home.nps.gov/applications/tuma/bibliography/ [accessed: May 23, 2013].

Griffith, J.S. 1992. *Beliefs and Holy Places: A Spiritual Geography of the Pimería Alta.* Tucson: University of Arizona Press.

Gomezcésar Hernández, I. 1997. Marco histórico de la migración Sonorense a Arizona. *Sonorenses en Arizona: Proceso de Formación de una Región Binacional,* edited by F. Lozano Ascencio. Hermosillo, Sonora, Mexico: Universidad de Sonora, 1–62.

Hackenberg, R.A. 1964. Aboriginal Land Use and Occupancy of the Papago Indians. As reprinted in *Papago Indians I* (1974), edited by D.A. Horr. New York: Garland Publishing, Inc., 23–308.

Hardwick, S.W. and Mansfield, G. 2009. Discourse, Identity, and "Homeland as Other" at the Borderlands. *Annals of the Association of American Geographers*, 99(2), 383–405.

Henderson, M.L. 1991. Duality in Modern Chiricahua Apache Settlement Patterns. *Traditional Dwellings and Settlements Review*, 2(2), 7–16.

Hoover, J.W. 1935. Generic Descent of the Papago Villages. *American Anthropologist*, 37(2), 257–64.

Johnson, C., Jones, R., Paasi, A., Amoore, L., Mountz, A., Salter, M., and Rumford, C. 2011. Interventions on Rethinking "the Border" in Border Studies. *Political Geography*, 30(6), 61–9.

Jones, R.D. 1969. An Analysis of Papago Communities 1900–1920. PhD Diss., Anthropology. Tucson: The University of Arizona.

Jones, R. 2012. Spaces of Refusal: Rethinking Sovereign Power and Resistance at the Border. *Annals of the Association of American Geographers*, 102(3), 685–99.

Joseph, A., Spicer, R.B., and Chesky, J. 1949. *The Desert People: A Study of the Papago Indians*. Chicago: University of Chicago Press.

Konrad, V. and Nicol, H.N. 2011. Border Culture, the Boundary between Canada and the United States of America, and the Advancement of Border Theory. *Geopolitics*, 16(1), 70–90.

Kelly, W.H. 1963. The Papago Indians of Arizona: A Population and Economic Study. As reprinted in *Papago Indians III* (1974), edited by D.A. Horr. New York: Garland Publishing Inc., 9–149.

Lewis, D.R. 1994. *Neither Wolf nor Dog: American Indians, Environment, and Agrarian Change*. New York: Oxford University Press.

Lopez, D. 2004. O'odham Culture and Heritage: T-Ni'oki (our language). *The Runner*, 23 June: 11.

Lopez, D., Reader, T., and Buseck, P. 2002. *Community Attitudes toward Traditional Tohono O'odham Foods*. Sells, Arizona: Tohono O'odham Community Action and Tohono O'odham Community College.

Luna-Firebaugh, E. 2005. 'Att hascu 'am 'i-oi? What Direction Should We Take? The Desert People's Approach to the Militarization of the Border. *Washington University Journal of Law and Policy*, 19, 339–63.

Madsen, K.D. 2005. A Nation across Nations: the Tohono O'odham and the U.S.–Mexico Border. PhD Diss., Geography. Tempe: Arizona State University.

Madsen, K.D. 2007. Local Impacts of the Balloon Effect of Border Law Enforcement. *Geopolitics*, 12(2), 280–98.

Madsen, K.D. 2014. The Alignment of Local Borders. *Territory, Politics, Governance*, 2(1), 52–71.

Manore, J.L. 2011. The Historical Erasure of an Indigenous Identity in the Borderlands: the Western Abenaki of Vermont, New Hampshire, and Quebec. *Journal of Borderlands Studies*, 26(2), 179–96.

Martínez, O.J. 1994. *Border People: Life and Society in the U.S.–Mexico Borderlands*. Tucson: University of Arizona Press.

Martinez, A.E. and Hardwick, S.W. (2009). Building Fences: Undocumented Immigration and Identity in a Small Border Town. *FOCUS on Geography*, 52(4), 48–55.

Meeks, E.V. 2003. The Tohono O'odham, Wage Labor, and Resistant Adaptation, 1900–1930. *Western Historical Quarterly*, 34(4), 468–89.

Meeks, E.V. 2007. *Border Citizens: The Making of Indians, Mexicans, and Anglos in Arizona*. Austin: University of Texas Press.

Megoran, N. 2004. The Critical Geopolitics of the Uzbekistan–Kyrgyzstan Ferghana Valley Boundary Dispute, 1999–2000. *Political Geography*, 23(6), 731–64.

Megoran, N. 2012. Rethinking the Study of International Boundaries: a Biography of the Kyrgyzstan–Uzbekistan Boundary. *Annals of the Association of American Geographers*, 102(2), 464–81.

Nabhan, G.P. 1982. *The Desert Smells like Rain: A Naturalist in Papago Indian Country*. New York: North Point Press.

Nabhan, G.P. 1985. *Gathering the Desert*. Tucson: University of Arizona Press.

Nagel, C.R. 2009. Rethinking Geographies of Assimilation. *The Professional Geographer*, 61(3), 400–407.

Naranjo Jr., R.V. 2011. Hua a'aga: Basket Stories from the Field, the Tohono O'odham Community of A:l Pi'ichkiñ, Sonora, Mexico. PhD Diss., Interdisciplinary Program in American Indian Studies. Tucson: The University of Arizona.

Newman, D. and Paasi, A. 1998. Fences and Neighbours in the Postmodern World: Boundary Narratives in Political Geography. *Progress in Human Geography*, 22(2), 186–207.

Nolasco, A.M. 1965. Los Pápagos, Habitantes del Desierto [The Papagos: Inhabitants of the Desert]. *Anales del Instituto Nacional Antropología e Historia* [Annals of the National Institute of Anthropology and History], 17, 375–448.

Oblasser, B. 1936. Papagueria, The Domain of the Papagos. *Arizona Historical Review*, 7(2), 3–9.

Paasi, A. 2009. Bounded Spaces in a "Borderless World": Border Studies, Power, and the Anatomy of Territory. *Journal of Power*, 2(2), 213–34.

Papago Indian Agency. c.1942. Place names on the Sells Gila Bend and San Xavier Indian Reservations. US Department of the Interior.

Prokkola, E.-K. 2009. Unfixing Borderland Identity: Border Performances and Narratives in the Construction of Self. *Journal of Borderlands Studies*, 24(3), 21–38.

Runner, The. 2003. Bid by O'odham in Mexico to Stop General Election is Turned Away by Court. 20 May: 1.

Saxton, D. and Saxton, L. (eds). 1973. *O'otham Hoho'ok A'agitha: Legends and Lore of the Papago and Pima Indians*. Tucson: University of Arizona Press.

Saxton, D., Saxton, L., and Enos, S. 1983. *Dictionary: Papago/PimaEnglish, O'othham-Mil-gahn; English-Papago/Pima, Mil-gahn-O'othham*, edited by R.L. Cherry. Second ed. Tucson: University of Arizona Press.

Schulze, J.M. 2008. Trans-nations: Indians, Imagined Communities and Border Realities in the Twentieth Century. PhD. Diss., History. Dallas: Southern Methodist University.

Spears, A.L. 2005. *Sociopolitical Boundaries and the Communication of Collective Identity on the Tohono O'odham Reservation*. MA thesis, Anthropology. Edmond: The University of Oklahoma.

Spicer, E.H. 1962. *Cycles of Conquest: The Impact of Spain, Mexico, and the United States on Indians of the Southwest*. Tucson: University of Arizona Press.

Singleton, S. 2008. "Not our Borders": Indigenous People and the Struggle to Maintain Shared Cultures and Polities in the Post-9/11 United States. *Journal of Borderlands Studies*, 23(3), 39–54.

Tohono O'odham Legislative Branch. 2013. Tohono O'odham Code, Title 12—Elections. Available at: http://www.tolc-nsn.org/docs/Title12Ch1.pdf [accessed: September 20, 2013].

Tohono O'odham Nation. 2013. Districts. Available at: http://www.tonation-nsn.gov/districts.aspx [accessed: September 6, 2013].

Toupal, R.S. 2001. Landscape Perceptions and Natural Resource Management: Finding the "Social" in the "Sciences." PhD diss., Landscape Architecture. Tucson: The University of Arizona.

Sheridan, T.E. 2006. *Landscapes of Fraud: Mission Tumacácori, the Baca Float, and the Betrayal of the O'odham*. Tucson: University of Arizona Press.

Strüver, A. 2005. "Everyone Creates One's Own Borders": The Dutch–German Borderland as Representation. *Geopolitics*, 9(3), 627–48.

Underhill, R.M. 1939. *Social Organization of the Papago Indians*. New York: Columbia University Press.

Van Otten, G.A. 2009. Illegal Immigration and Smuggling on the Tohono O'odham Reservation of Arizona: Opportunities for Applied Intelligence Training. *Military Intelligence Professional Bulletin*, 25(2), 38–45. Available at: http://www.fas.org/irp/agency/army/mipb/2009_02.pdf [accessed: September 9, 2013].

Waddell, J.O. 1969. *Papago Indians at Work*. Tucson: University of Arizona Press.

Weir, D.R. and Azary, I. 2001. Quitovac Oasis: A Sense of Home Place and the Development of Water Resources. *Professional Geographer*, 53(1), 45–55.

Wilbur-Cruce, E.A. 1987. *A Beautiful, Cruel Country*. Tucson: University of Arizona Press.

Wilson, C.R. 1972. Migration, Change, and Variation: A Papago Case Study. PhD Diss., Anthropology. Boulder: The University of Colorado.

Chapter 6

Whose Border? Border Talk and Discursive Governance of the Salween River-Border

Vanessa Lamb

Introduction

A coalition of developers in Thailand, China, and Burma are working to gain political approval for the Hatgyi dam, one of 16 large hydroelectric projects proposed on the transboundary Salween River. The dam is proposed in Burmese territory, with anticipated impacts in Burma as well as upstream in Thailand and China. If built, Hatgyi would be the first dam on this river, one of the longest free-flowing rivers in Asia, threatening the livelihood and food source of the six million people who live in the river basin (Wong et al. 2007). The dam also poses challenges for the river's political geographies. Within a short distance upstream of the Hatgyi dam site, the river comprises 120 kilometers of the political border that separates Thailand and Burma (see map, Figure 6.1). The proposed construction of the dam and the contemplated water management schemes associated with its installation will transform the river's flow and water levels. As a consequence, a whole host of questions have emerged from activists, government officials, border residents, and military officers about how these physical changes will affect the political border.

While these questions and concerns focus on the challenges to the border posed by the dam's construction, I argue that these miss a significant element of the way that the political border is transformed. Instead, the planning and participatory development processes of Hatgyi dam are reconfiguring the political border before the actual construction of the dam. This is a consequence of discursive work by the proponents, government agents, villagers, and activists to remake the political border for specific aims and particular narratives and histories, in order to influence the project's decision-making processes.

My research builds on data generated through an ethnographic approach to the making of borders in everyday life. Based on this research conducted at the Salween River where it comprises the Thai–Burma border in 2010–11, I argue that "border talk" is an overlooked part of the process of shaping questions about who borders and how (Johnson et al. 2011). This approach illustrates that the border is not simply imposed nor is it a "natural" line that development crosses, but the border is continually rearticulated in the processes of cross-border resource development.

**Figure 6.1   Map of the Salween River-border indicating location of
the proposed Hatgyi hydroelectric project and study site**

Note: Original cartography by Carolyn King, York University.

For instance, in interviews and meetings in Thailand many residents
emphasized that, in addition to the impacts on livelihood and food, what was at
stake for them in the project was the loss of their understanding and knowledge
of the area and of the river-border in particular. As one local leader said, many
residents are concerned about their security and about the river-border's place
in their lives, questioning "will it be the same as before?" In this, he highlighted
the uncertainty about whether the river-border area would go back to the way it
was 30 years ago, when these communities considered themselves more "cut off"
from Thailand and saw their lives as "less secure."

Because the border is such a fraught political idea, calls to take note of and
secure the border are a way for residents and other actors to claim a place in
decision-making from which they might otherwise be excluded. At the same time,
official talk about the border can serve to restrict the discursive space for the

same residents whose compliance these decision-making processes seek to enroll. Overall, I argue that the ways that people talk about the border, as part of a larger project of discursive governance, are significant in that they both influence and open up possibilities for how the political border can be remade.

Talk of Borders

As part of a long-standing conversation in Southeast Asia about the creation of political borders and nation-states (Winichakul 1994, Leach 1960, Rajah 1990, Wijeyewardene 1991, Grundy-Warr and Wong 2002, Duncan 2004, Dean 2005, Walker 1999, Baird 2010, Sturgeon 2004, Scott 1998, 2009, Horstmann 2011), Peter Kunstadter argued that "the present political borders in Southeast Asia represent a series of events" (1967: 18). His work productively intersects with recent arguments by border scholars that borders are not lines but represent spatial and temporal processes, that can be remade, resisted, or undone through practice and performance (Norman 2012, Jones 2012, Salter 2010, 2011, Doevenspeck 2011, Prokkola 2008, 2009, Rumford 2008, Doty 2007, Balibar 2002, Newman and Paasi 1998, Paasi 1996; see also Mountz 2010 on state as performance).

Building on this work, my findings highlight how the making and remaking of these borders, through "border talk," includes (but is not limited to) the very communities and individuals that reside at the border. This complements the assessments of Kunstadter and other scholars of Southeast Asia (i.e. Leach 1960, Winichakul 1994), which posit that border events were the work of "outsiders"—western colonizers and central governments—imposed on local populations living at, around, and across those borders. Through analysis of border talk, I address how the border is presently positioned and enacted in the lives and narratives of border residents, and I also demonstrate that what is considered the "border" shifts over time and place for a variety of actors.

The concept of border talk has been employed in existing literature to elucidate shifting identities (i.e. Pickering 2006, Friedman 2002; see also Doevenspeck 2011 who examines border talk as part of border narratives at the Congo–Rwanda border). Building on this scholarship, which highlights the multifaceted and shifting identities of borderlanders as articulated through their own border talk, my analysis focuses on how the *border* is positioned, repositioned, and put at stake in and through the words, lives, and narratives of a variety of actors. Border talk represents part of the discursive governance of the political border; it both influences the way that the political border is remade and also opens up possibilities for how the political border is remade or undone. In other words, it shapes what is possible and what is less possible. What I highlight through instances of border talk at the Salween is how these articulations of the border are part of shaping nature–state–society relationships.

In particular, I address the implications of border talk for the political and the sub-political relationships between the border and its residents, authorities, developers, and activists. In the context of this case at the Salween River-border, the policing of citizenship, the management of cultural and national identities, and the maintenance of local structures of authority and resource distribution are complicated by the imposed specter of large-scale river development.

I also position this case as useful for thinking about the conflated institutional and ecological functions of the river-border and the ways that nature-state-society relationships are forged through border talk. Rivers "used" as borders number approximately 200 globally (Varis et al. 2008: 6–7; see also Fall 2010). In this case where the river comprises the political border, it is conceivably more obvious how the political border is at stake in conversations because of concerns that changes to the river will have very real material implications for the border. At the same time, talking about the border when it is also a river makes discussion of the border more complicated. It may obscure an acknowledgement of the border within economic, ecological, or geophysical language. To address these complications, I also pay attention to gaps, absences, or silences as part of border talk. Variations in representations of the river, the border, and their common position can reveal relationships among those speaking (or not speaking as the case may be), and between individuals and the border, the nation-state, and nature.

To make these arguments, I draw on data generated through 12 months of multi-sited ethnographic fieldwork, both at and away from the border. Following Megoran's (2006) call for an ethnographically informed political geography, this chapter draws on interviews and participant observation carried out with over 100 individuals including border residents, activists, developers, and government officials in Thailand from September 2010 to August 2011. I conducted a multi-site ethnography, tracing the unfolding development planning process of the Hatgyi dam in the daily life of the villages at the river-border, but also in the meetings of government and activists at the border and at other locations in Thailand. This research and analysis also draw from preliminary fieldwork conducted in 2009 and work experience in the region over the past decade.

One of the ways that I present the "border talk" of multiple actors is through explication of the dialogue between residents, activists, government officials, and military soldiers at the dam's public information hearing. Here, the dialogue simultaneously invokes visions and experiences of a porous border and of one that is well-demarcated and secure; I pay attention to how speakers' operational roles drive the ideas, images and concerns that they raise at the meetings, as well as their places within that dialogue. Before I present this ethnographic data, I first introduce my site of research in order to provide a context for the border talk of residents, government officials and others.

Histories of the Border: Placing the Salween River-border in Context

The Salween River comprises approximately 120 kilometers of the political border between Thailand and Burma (see map, Figure 6.1). According to Thongchai

Winichakul, the Salween River-border was demarcated in 1849 (1994: 69) and was negotiated between the British administration in Burma and the northern kingdom of Chiangmai, part of present-day Thailand. The negotiations were conducted by the British in order to secure their claim to teak timber on the west bank of the Salween. That this site for the border was chosen and then delimited according to the British interest in teak, highlights the intersection of the historical development of natural resources as tied to territory-making (see also Vandergeest and Peluso 1995). This means that the planned site of the Hatgyi dam is part of present-day Burma.

More broadly, Winichakul (1994) argued that such negotiations with European colonizers transformed "indigenous" understandings of the border, for whom it was overlapping and ambiguous, to represent the colonizer's Cartesian understanding of the border as a fixed line on a map. Interestingly, Winichakul (1994: 69) noted that in the case of the Salween:

> Within two years, [the British] had surveyed every fork of the Salween River in order to identify the main stream which would be regarded as the boundary. And with the assistance of the five oldest Karen along the river, they finished the job of marking the modern-style boundaries in 1849.

This description is one of multiple historical works which draw attention to the Karen people residing at the river prior to the making of the present-day "modern" boundary (see also Wilson and Hanks 1985). In this instance, Winichakul also highlights how their knowledge and experience at the river was invoked to assist in the fixing of the boundary.

Today, a large number of people residing here identify as Karen, and the landscape of the Salween River-border still shows evidence of its timber-producing history. Rows of planted teak trees still mark multiple points along the river where the land was used by various companies and by the Thai state for timber extraction, even though timber is no longer legally extracted in this area since the 1989 logging ban in Thailand.[1]

Because the river is a border, much of the everyday riverine livelihood activities are entangled in the institutional functions and maintenance of the border. For instance, the river-border is policed from both banks. In order to travel between two villages on the Thai side of the river, boats heading either up or downstream from the main pier (located at Mae Sam Lap, see Figure 6.1) may pass through a number of checkpoints on either side of the river. Thai border guards, the Karen National Union (KNU), and the Democratic Karen Buddhist Army (DKBA) under the Burmese Border Guard Force all maintain intermittent check points. In addition, boat traffic is only allowed on the river from 8 or 9 am until 5 pm. Traders, fisherpersons, and tourist or other transport launches, are not

---

1   Illegal timber harvests continue, but do not approach the scale of past forestry. Old teak plantations now represent outposts of state regulation along the river, off-limits to local and regional economic activity.

allowed to be on the river outside of these hours. This river-border, with all of its expectations, constraints and violations, is now at the center of discussion in the decision-making processes of the Hatgyi hydroelectric dam proposal.

The project, if it goes ahead, would be a large 1200 megawatt hydroelectric dam located near the political border just where the river flows into Karen State, Burma (see Figure 6.1). Ninety percent of the electricity would be sold to Thailand; the remaining 10 percent is to remain in Burma. The Memorandum of Understanding to develop the Hatgyi project was agreed to and signed by the international arm of the Electricity Generating Authority of Thailand (EGAT), along with China's Sinohydro Company and two Burmese investors: Burma's Ministry of Hydropower, and IGE, a privately held Burmese company.

Among the many actors involved in this development, Thailand's state electrical authority has perhaps the most at stake in the way that the political border is defined. Part of the debate over whether this project will proceed focuses on whether the dam, a short distance downstream of the length of the river that serves as the international boundary, will flood that border. Following Article 190 of the Thai constitution, any project that proposes to disrupt the country's political border requires parliamentary approval. Questions raised about the dam's impacts to the political border resulted in repeated modification and extension of the decision making process that must precede the development. The 2009 establishment by the Thai Prime Minister's Office of a Hatgyi Subcommittee occurred largely as a result of continuing concerns about the development's cross-border effects. The subcommittee was mandated to address those questions; in 2011 it organized public information hearings meant to provide project details to stakeholders in three Thai districts adjacent to the river-border.

Some of the key examples of border talk that I describe in this chapter occurred during these hearings. In these public information hearings, residents emphasized the significance of the river-border to their everyday activities, as well as to the position and security of their villages.

The potential for economic hardship and population displacements has driven project developers, anti-dam activists and border residents to produce a host of claims and representations about the impacts that will accrue from the dam. An important component of how these claims about the Karen are made by all sides in the debate relates to their position in Thailand as an ethnic minority group. In Thailand, many ethnic minorities, but particularly the Karen, are often positioned as living close to nature and far from modernity (Forsyth and Walker 2008, Delang 2003). In these stereotypes and narratives, the Karen are responsible for taking care of "nature," but they are also blamed for any deforestation and other environmental degradation that occurs (Forsyth and Walker 2008, Walker 2003). The popular accounts of the Karen, not only in media and historical scholarship but also in environmental campaigns, also place them as living at the political-cultural edges of Thailand, far from those who would be considered to be politically or ethnically "Thai" (Winichakul 2000, Vandergeest 2003, Keyes 2003, Forsyth 2007, Forsyth and Walker 2008). Instead, they are more frequently associated with the Burmese nation-state.

This contested positioning of Karen people intensifies the discussion of what is at stake when talking about the border, about which side of the border residents belong (even when living in villages with long standing within Thailand), and about whose jurisdiction they may represent. Border talk is important because it facilitates analysis of the ways that the border is invoked by residents and other actors, revealing contrast between different individuals' articulations of the political border and helping us to understand their contributions to development outcomes.

In the next three sections, I illustrate examples of border talk by residents, military officers, activists, and government officials. The aim of presenting this range of border talk is to highlight precisely how the border's meaning shifts through the statements of multiple stakeholders in the cross-border development process.

Developing the Border: The Hatgyi Subcommittee and Environmental Governance

Border talk was an integral part of the Hatgyi public information hearings which represented an opportunity for exchange between residents, officials, soldiers, and activists. The hearings were held in early 2011, and the meeting that I draw from here was held in one district capital in northwestern Thailand adjacent to the river-border. The mandate for these hearings was to present in the words of one subcommittee member "the truth" about the project, particularly on the question of whether or not the dam would inundate parts of Thailand, or would otherwise impact the political border. The border talk in these meetings demonstrates the discursive work of multiple actors to remake the border in ways intended to either facilitate the development and to argue against its approval. In turn, their discussions highlight how articulations of the "border" shift.

The Hatgyi subcommittee was composed of 19 members, including EGAT officials, Ministry officials, the environmental consultant who carried out the EIA, five civil society representatives including NGO staffers and a member of the National Human Rights Commission of Thailand, and a military officer named Khet.[2] The audience was mostly comprised of village headmen and residents from affected districts in northwestern Thailand. In addressing the audience, Khet referred to the project as an opportunity to finish what the British started in the late 1800s: marking out the "official" border. He explained that in the past, "They [Siam and British] decided that Thai would have one bank, and the British the other ... but in the present, ... there are only 500 kilometers that are clear; there are still 2,400 kilometers that have not yet been surveyed. This is just the beginning..." (Feb 9, 2011). According to his presentation, cross-border development and official demarcation of the border were directly linked; the development project, whatever its impacts, represented an unprecedented opportunity to improve the

---

2   All names used for interviewees are pseudonyms.

fixity and order of the existing border. EGAT officials and the environmental consultants they hired also presented their reassurances that the dam project would not inundate the political border. One of the presenters frankly stated that those concerned about flooding or the political border "should not worry" (hearing transcript February 9, 2011), underlining the message that the project did not pose significant consequences for the border.

There were no presentations by village representatives. There was an opportunity for residents to ask questions and many did so, raising issues affecting the livelihood activities and travel on the river. The relationship of the border to boat traffic dominated this discussion, with audience members inquiring about exactly where (and when) each country could exercise its authority over the river. For instance, one man, who did not introduce himself but I recognized from a village along the Salween that is known for its part in cross-border trade, asked, "Who has the authority to close the river to boat traffic—Burma or Thailand"? (hearing transcript February 9, 2011)

Instead of addressing the more expected issue of fishing impacts, residents raised questions about travel and transport. Within the ensuing discussion, the underlying context of public discussions at the hearings was related to questions of control and authority—over the river, the border, and the people living on each side. The timing of the hearing is also important here—these hearings took place during a period in 2010–11 in which official and unofficial "closures" of the river-border occurred due to civil conflict in Karen State, Burma, a preoccupation for many whose livelihoods depended on river traffic. These discussions were also spurred by broader, more long-standing connections between an orderly border and the political and legal legitimacy of village residents, and by Khet's comments on the development's association with border demarcation and security.

The Chair of the Subcommittee repeated the assertion that this project would lead to clarification of these border issues. Then, in his haste to conclude discussion of boat traffic, he addressed the audience:

> Do people understand who is responsible? I am sure if [Burma] was without conflict, then this would not be an issue … with regard to what happens with the boats, I do not understand the question exactly. … If you are asking why the Burmese government can prohibit Thai boats from travelling on the river, it can't … We share the river. (February 9, 2011, field notes)

The Chair's comments served to limit further discussion on this topic. He later reiterated that the border and "Thai soil" would not be affected and were not of primary concern, maintaining that EGAT was more than capable of making sure that the political border would not be altered.

As was evidenced through the public information hearing, it was not necessarily accepted that Hatgyi dam could be constructed so close to the political border and not impact Thailand, but this is exactly what must be accepted for the development to go ahead without becoming a subject of national political debate.

In a move to draw attention to information not presented at the public information hearing, NGO activists from the local district and from farther afield expressed shared concerns about the dam, the border and security. Towards the end of the hearing, they read aloud letters and asked pointed questions about the (conflicting) evidence for border impacts documented in a Human Rights Commission of Thailand report (NHRCT 2000).[3]

One activist named Patana highlighted what he considered to be the cross-border impacts of the project that were not addressed by the Subcommittee's presentations. He told the meeting room, "For those who are not based at the river, I want you to see the kinds of impacts you will get." He raised the issue of human rights abuses, known to be perpetrated by the Burmese military against residents in Karen State. Patana declared that, in contrast to the democracy touted in the context of the 2010 November elections in Burma, human rights abuses in Karen State near to Thailand were increasing. He concluded his list of concerns with a warning about the potential for future problems, appealing to the Hatgyi Subcommittee about the problems for the Karen people in Burma:

> And for the Karen who opposed the dam, with the increase in military [Burmese soldiers] in the area, there will be more refugees. Now, there are already refugee camps set up. They are full. What will happen when the dam is built and more refugees come? Where will they go? Can you tell me—what will the impacts be? The people in the camp will leave the camp or they won't be able to stay in the camp because it is full. (February 9, 2011)

Patana directly linked the Hatgyi dam with issues of border security, with armed violence in Burma, and with the issue of refugees crossing the border into Thailand. His talk of the border shows two different discursive moves: the first was to call attention to the fact that this public information hearing in Thailand was also about Burma. This was one of the first times I witnessed the issue of impacts within Burma raised at these meetings; it was mostly overlooked up to this point, helpfully so for a Thai subcommittee with no authority to answer on behalf of Burmese officials, and of residents and officials in Karen state. Second, the activist highlighted interlinked concerns about refugees, illegal migrant workers, and the dam's potential to "unmake" the political border by spurring additional movements of people across it. The activist's suggestion of a cross-border spread of impacts from within Burma resonated in interesting ways with Thai-side residents' own concerns about the unmaking of their relationships with the river-border.

Patana, as an advocate for human and environmental rights, was attempting to make sure these concerns about violence, the plight of refugees, and human rights abuses more generally were included in the hearing record. Yet, his words also

---

3   This report show that cross-border impacts would be expected from the Hatgyi dam and was part of the impetus for the Prime Minister's office initially establishing the Hatgyi Subcommittee.

conjured emotions, in particular a broadly held set of fears about border security, and connected them with the Hatgyi project.[4] At the conclusion of Patana's rather long "question," the Chair discouraged other subcommittee members from responding, and urged the "real" stakeholders to produce questions for the subcommittee.

Will it be the Same as Before? Changing Articulations of the River-border

At the dam's public hearings, residents mainly claimed discursive space to ask questions about issues of control and authority of the river-border. However, as I detail below, outside of the public information hearing, talk of the border and the state were also manifested, although in more discrete ways.

In one of the three villages where I conducted fieldwork along the Salween River-border, I interviewed Wandii, and her husband, Wanankan. They live in Thailand's Mae Hong Son Province just upstream from the proposed Hatgyi dam site. Their village is comprised of about 500 households. Most individuals identify as Karen or Thai-Karen ethnicity as Wandii and Wanankan do, and while the majority of residents were born in Thailand, which would in many cases make them citizens, but many do not have birth or citizenship documentation. The village is not included on most government maps and until recently was not marked on international digital mapping products. For these reasons, many activists refer to these villages as "nokpantii" or "off the map," and local NGOs are working to map and count villages and residents with the aim of helping individuals apply for citizenship documentation and forcing development activities to acknowledge the villagers' legal presence and their potential to experience project impacts.

Wandii described life in the village as not very good, and said that she would welcome development that would bring decent jobs to her village. Growing up along the Salween River-border her entire life, at 46 years old Wandii saw many changes. She explained that "there was an increase in the number of people and an increase in village development," of which she noted a new road and clinic. But in her opinion, the quality and quantity of agricultural activities decreased. She made clear that "It used to be fertile, there used to be enough to eat. But now there is not enough." She reiterated that she wants her children who attend school in another town to come back to the village, but she is sure that there is not enough space for them to participate in agriculture or to build a house, and that it would be difficult for them to make a living. This year, their household only received 60 tang [local measurement equivalent to one large bucket] of rice. It was full of insects and was not enough to feed her family. She asserted that, "Even as we see more development, the agriculture and food for us is not so good." When we discussed the proposed dam, Wandii was neither adamantly against nor for the project.

4　Problematically, this talk of "outsiders" was picked up and reiterated throughout the hearing.

For many like Wandii, the proposed cross-border development might offer some benefits, but is not necessarily beneficial overall to the existing livelihood and income-earning activities of local residents. The project, on its own and as one of 16 proposed dams, may also necessitate relocation in the longer term—not necessarily due to the flooding of resident's homes but as a result of more insidious social displacements that can accompany large-scale dam development (Vandergeest et al. 2007). Her concerns are also related to a larger tension where many individuals are participating in state-led development processes because of potential benefits, but are also critical or even skeptical of the stated motives of the process or its outcomes. Wandii's experience with development resonates with critiques of the participatory turn in development procedures (Ferguson 1994, Käkönen and Hirsch 2009), and illustrates how individuals consider and grapple with the opportunities presented through their own or their village's participation in development processes.

While development and the impacts to the river and daily life were always emphasized, what I was struck by in interviews with residents like Wandii at the river-border were the silences regarding the political border itself. The border was not talked about in a direct sense. Instead, it was the "river," "development" and the act of "crossing" that residents directly talked about in village meetings and in interviews. Yet, even as the political border was not named in our discussions, facets of the political border and residents' changing relationships with the river border were articulated as part of everyday life.

This could be seen, for instance, in border residents' use of storytelling to contextualize contemporary events and concerns. One local leader, named Thomas, repeatedly raised a saying he identified with his ancestors: "If the Salween runs dry, we cannot come and go [visit one another]. To cross the river would be a sin, even if we try to cross, we cannot." (Thomas, October 17, 2010) While I first heard this saying from Thomas, I heard it in different variations by at least 10 others in the same village. The main point communicated here is that changes to the river flow would transform the presently crossable, negotiable river into an impermeable border. According to Thomas' ancestors, crossing would bring significant undesirable but unknowable consequences for the village.[5]

While it is difficult to imagine the river-border as it is now—a rather turbulent flow of water that even during the low season provides water—transformed to a dry river bed, for Thomas and others in the village the potential for dam development to change the river's flow regime presents the possibility that the river will be similarly hardened and made unavailable to them. In discussions, the

---

5   To highlight the significance of this story further, there are other times when it is not advised to cross the border, such as when the river "runs red" (which means it is the high season and the river is quite turbulent), but in comparison, those "not advised" occasions are different from the river running dry. The dry riverbed represents a significant event and it actually constitutes a "sin" to cross it.

proposed dam was repeatedly linked to this vision of the river running dry and to its transformation from a site of "life" to a barrier.[6]

Even with the unlikelihood of the river running dry, the discursive shift is an important one. Whether Thomas presently crosses the border with much frequency, this change from "being able to" to "cannot" cross is significant. Although he stressed disturbing the natural river flows, there are also implications for not just the ecological functions of the river, but also the institutional functions of the border. In practical terms, it signifies that not only would the river would dry up but that trading and selling along and across the river-border would also be impacted. In this case, trade is made lucrative because of the border. Were this river not a border, with the associated complications and opportunities for trade and transport that it offers, concerns might be articulated differently, and would be less obviously significant to the everyday lives of residents. The river-border is also used as a personal transportation route for most villages along it, which lack all-season road connections to each other and to district centers. As such, the specter of the river's transformation is enhanced for those with family members or relationships with individuals living in other villages.

This discursive shift from a site of life to a barrier is also significant within the village as a way to think about a future of ecological and institutional uncertainties. The specter of river-development for the village is not just related to Hatgyi, but to all the dams that are proposed upstream (some under construction now) in China, developments on which Thai and Burmese villagers did not receive any consultation. For border residents, the more managed and constrained flow-regime that will result from these upstream dams will transform the river and the livelihood opportunities it provides them. The dynamic, seasonal river that currently exists supports a variety of transitional fisheries as well as productive dry-season gardening of the river's exposed banks; its unpredictability also limits access to the river for conventional policing. A stable flow would alter the possibilities that current seasonal changes afford. In particular, stabilizing the river's flows year-round will affect and constrain villager livelihoods just as definitively as if the waterway "dried up"—both the flat river and the dry river are barriers to local lives. In my analysis, this discursive shift from "river as life" to "river as barrier" is not only directly significant to trade and subsistence futures, but also to thinking about the possibilities of the political border and of its implications for border residents.

In addition to these narratives which highlight the development at the border, Karen residents also presented and documented their histories as related to trees and timber as a precursor to the present dam proposals. Some of these histories

---

6   The community is upstream of the dam and I do not understand how the dam could cause the river above it to "run dry." However, I understand that the essence of the concern is that there will be a large change, particularly with regard to river flow, and this change will affect their lives in negative ways, which is consistent with Hatgyi project proposal and perhaps all the more relevant in the context of 16 proposed projects.

were recorded and textualized in local knowledge projects such as in the *Salween Study* (FER 2007). Residents explained how they associated the building of the dam with erasing their livelihoods, such as riverbank gardening and swidden agriculture, as well as residents' knowledge and history within Thailand. Many individuals reiterated the significance of making, documenting, and sharing their knowledge about the area, because as one elder put it, "If the dam is built, then all of these things disappear" (Dec 2010 fieldwork notes). Many expressed concern about their security and the river-border's place in their lives, including the local leader quoted in the introduction to this chapter, who wanted to know "will it be the same as before?" As noted, his words call attention to the uncertainty about whether the river-border area would go back to the way it was in the past, which was seen as "less secure." A main thread through these discussions and presentations of local history was that the relationships forged with the border from the Karen residents' position in Thailand was an important accomplishment.

A Soldier's Words: Concern for Thai-*land*

> We will only lose 100 rai of Thai soil—people can just move up [into the mountains]. (Thai Soldier, October 2010)

Outside of the public information hearings, I also informally interviewed a Thai soldier posted at the river-border while I was waiting for a boat on the Salween. This boat trip came after conducting prolonged fieldwork in Wandii's village along the river-border. The soldier, "Dan," recognized concerns about losing "Thai soil," but also expressed optimism about these impacts and what they would mean from the perspective of a state agent. In his comment quoted above, he references both the soil and territory, and the marginal loss of territorial area elides the much greater loss of intensely productive riverbanks accessible to village residents.

The soldier explained that the concession of 100 rai (approximately 16 hectares/39.5 acres) was acceptable because it would only affect gardens along the banks of the Salween River and the "minority groups" who maintain them. He contrasted this "lost soil" with references to his duty to protect the forests, national buildings, village temple, and connecting roads, aspects of the border landscape loaded with much greater official weight.

In addition to his use of the words "Thai soil" to refer to the Thai nation-state's claims to land, he also discussed ongoing "illegal Burmese timber" at the border. Such characterization of logging not only maps the 1989 Thai logging ban onto Burmese territory (where teak logging is not illegal), it was also the only time during our conversation that the soldier referred to Burma or to Burmese authority at the border.

I raise Dan's talk of "Thai soil" and "illegal Burmese timber" to highlight the often conflated political and ecological articulations of and at the border. In these conversations, the notion that the soil has a nationality facilitates particular conversations and can serve to exclude certain peoples. This element of exclusion is further highlighted in his above comments about ethnic minorities in Thailand.

Dan's border talk is similar to issues raised in interviews with dam developers, environmental consultants, and government officials who served on the subcommittee, and in their public comments at the hearings—particularly by the Chair who also invoked the term "Thai soil." How this "border talk" matters and what it means for the political border is part of what I consider next, in discussion.

Discussion and Conclusion: Border Talk and Discursive Governance

The accounts of the Salween River-border presented through the border talk of various actors show how the border is understood and positioned in and through different embodied relationships with institutions and ecologies. The focus on "border talk" highlights the fluid and sometimes ephemeral character of the border and the ways it is invoked—not only as a fixed line, but as emergent, dynamic, and multifaceted.

Using a similar approach, Doevenspeck examined border talk as integral to narratives of the Congo–Rwanda border. He argued that "The range of accounts of the border, from being a desired barrier against the demonized 'other' and a means of exclusion to its conception as a resource, makes it clear that understanding the border through narratives requires allowing for historicity, relationality and contrariness" and "providing insights into state-society relations" (2011: 140). What this case adds to these insights is the intertwined production of nature–society–state relationships that occurs during development activities. For instance, the concerns expressed by activists and government officials regarding "illegal migrant workers" and the dam's potential to "unmake" the political border in ways that would increase illegal border crossing appeal to not only state–society relationships, but to visions of nature and nature's development. Added to these statements in public hearings, Dan's statements regarding "Thai soil" and minority groups at the border emphasize that he knows "whose border" it is. His comments combined with the concerns of border security and demarcation in the public information underscore how redefining the border through exclusion is a strong thread through this cross-border development.

These instances of border talk signify the redefinition of relationships with the political border. As the border's relationships with particular visions of security are debated, there are implications for the kinds of "border talk" and the relationships that are subsequently articulated. In this case, there is evidence of a tightening of discursive space for Karen residents who would be affected by the dam to present on and speak of the border. This is evidenced in the ways that Karen residents are expected and even encouraged to speak about topics related to their livelihoods and nature, but then are not asked about or encouraged to speak about the political border. It also influences the ways that individuals position themselves in relationship to that border within public meetings about the project. This border talk matters because it shapes the sets of issues and relationships invoked through the broader processes of cross-border development.

However, even as Karen residents do not name the border, outside the hearings they situate the river-border within their local historical narratives, and also located themselves as part of Thai historical narrative by delineating their role—past and present—as related to the Thai-land and to natural resources development. Residents took the opportunity presented by the development process to articulate claims to and about the river-border, connections which are not only forged through cross-border development processes but are also made more visible through such processes. This includes instances such as the Karen better able to speak about riverine livelihood activities and their important historical relationships with the river-border as part of local knowledge and also in questions posed at the public information hearing. I argue that even though the "state" and the "border" are not named, they are in many ways brought into conversations about development and everyday life, and both reflect and shape residents' relationships to the river and to the border of the Thai nation-state. The points raised here contribute to the border literature in Southeast Asia, which envisions the political border as "process" but does not locate minority groups as actors within those processes.

Moreover, such border talk highlights how these connections with the border are continually expressed and remade. These relationships, in turn, build on narratives (both historically produced and manifested contemporarily) about development and about the Karen that are important to consider, particularly in terms of the challenges to transform the way that borders, nature, and security are talked about by individuals at and away from the border.

On the one hand, these relationships are expressed in ways that build on problematic national and global discourses about immigration which position those near or at the border in a similar category to border crossers as national security threats. In this particular case, this positioning builds on a deeper history of contention between Thailand and Burma as well as on Thailand's history of defining the nation ("Thai-ness") through exclusion (Winichakul 1994, 2000). This framing is at least partially defined in conversation with histories and security discourses which already delimit the Thai-land and limited the space for the speech of border residents and "migrants."

On the other hand, the Chair and soldier asserted in the hearing that the border would be more solidly demarcated and secured through development. These examples draw attention to the ways that the project is invoked as fixing the border, in addition to (or in contrast to) the dam's "unmaking" of the border through its perceived impacts to increase illegal border crossing. The strength of border talk in this instance is that it can evince both sides of these contradictory articulations of the border, as related to the inclusions and exclusions of security discourse.

In addition, the border talk around Hatgyi not only invoked said visions of security, it also invoked onerous nature/nation constructions that included "Thai soil" and "Burmese illegal logging." Considering that this discourse on security emerged throughout the development process, the relationships forged through these words appeal to problematic apolitical or Malthusian approaches to understanding ecology and politics which position local resource users as problems for the state and for development to manage and "solve." In particular, this border talk echoes Fall's critique of the problematic links made to "natural" borders like rivers, which are mobilized in the name of naturalizing problematic assumptions that states should be ethnically homogeneous (2010: 140). Examining border talk reveals that these articulations of the border are worthy of scrutiny, and furthers understanding of the ways that particular understandings of the border are produced or "naturalized." The relationships forged through cross-border development processes—and the questions raised about the border and development within these processes—comprise an important element of this research and the arguments presented here.

In sum, while cross-border resource development proceeds on the basis that the political border is a given or even "natural" line that development and decision-making must *cross*, I argue that it is actually the border that is re-articulated and remade in these processes. One way that this commonplace border talk matters is in how it shapes, limits, influences, and potentially opens up possibilities for different articulations of the border. It concerns the redefinition of Thailand's border with Burma, and residents' relationship to both the river (nature) and the border (nation-state). Relating these arguments to broader debates about whether the nation-state and its borders are increasing or decreasing in importance, this chapter points to the ways that borders are remade through everyday state-making performances, including those that afford substantial discursive space to multiple local and non-local actors.

References

Baird, I. 2010. Making Spaces: The Ethnic Brao People and the International Border between Laos and Cambodia. *Geoforum*, 41(2), 271–81.

Balibar, E. 2002. What is a Border? *Politics and the Other Scene*, edited by E. Balibar. London: Verso, 75–86.

Das, V. and Poole, D. (eds). 2004. *Anthropology in the Margins of the State*. Santa Fe: School of American Research Press.

Dean, K. 2005. Spaces and Territorialities on the Sino-Burmese Boundary: China, Burma and the Kachin. *Political Geography*, 24(7), 808–30.

Delang, C. (ed.). 2003. *Living at the Edge of Thai Society: The Karen in the Highlands of Northern Thailand*. London: Routledge.

Doevenspeck, M. 2011. Constructing the Border from Below: Narratives from the Congolese-Rwandan State Boundary. *Political Geography*, 30(3), 129–42.

Doty, R.L. 2007. States of Exception on the Mexico–U.S. Border: Security, "Decisions," and Civilian Border Patrols. *International Political Sociology*, 1(2), 113–37.

Fall, J. 2010. Artificial States? On the Enduring Geographical Myth of Natural Borders. *Political Geography*, 29(3), 140–47.

FER (Foundation for Ecological Recovery). 2007. Salween Study—Salween: Source of Life and Livelihoods. Bangkok: Foundation for Ecological Recovery [in Thai, Introduction in English].

Ferguson, J. 1994. *The Anti-Politics Machine: Development, Depoliticization, and Bureaucratic Power in Lesotho*. Minneapolis: University of Minnesota Press.

Forsyth, T. 2007. Are Environmental Social Movements Socially Exclusive? An Historical Study from Thailand. *World Development*, 35(12), 2110–30.

Forsyth, T. and Walker, A. 2008. Forest Guardians, Forest Destroyers: the Politics of Environmental Knowledge in Northern Thailand. Seattle: University of Washington Press.

Friedman, S. 2002. "Border Talk," Hybridity, and Performativity Cultural Theory and Identity in the Spaces between Difference. *Eurozine*. Available at: www.eurozine.com/articles/2002–06–07-friedman-en.html [accessed: Sept 27, 2013].

Grundy-Warr, C. and Wong, E. 2002. Geographies of Displacement: the Karenni and the Shan across the Myanmar–Thailand Border. *Singapore Journal of Tropical Geography*, 23(1), 93–122.

Horstmann, A. 2011. Sacred Spaces of Karen Refugees and Humanitarian Aid across the Thailand-Burma Border. *ASEAS*, 4(2), 254–72.

Jones, R. 2012. Spaces of Refusal: Rethinking Sovereign Power and Resistance at the Border. *Annals of the Association of American Geographers*, 102(3): 685–99.

Johnson, C., Jones, R., Paasi, A., Amoore, L., Mountz, A., Salter, M., and Rumford, C. 2011. Interventions on Rethinking "the Border" in Border Studies. *Political Geography*, 30(2), 61–9.

Käkönen, M. and Hirsch, P. 2009. The Anti-politics of Mekong Knowledge Production. *Contested Waterscapes in the Mekong Region: Hydropower, Livelihoods and Governance*, edited by F. Molle, T. Foran, and M. Käkönen. Chiang Mai: Earthscan, 569–77.

Keyes, C.F. 2003. The Politics of "Karen-ness" in Thailand. *Living at the Edge of Thai Society: The Karen in the Highlands of Northern Thailand*, edited by C. Delang. London: Routledge, 210–18.

Kunstadter, P. 1967. *Southeast Asian Tribes, Minorities, and Nations*. Princeton University Press.

Leach, E.R. 1960. The Frontiers of "Burma." *Comparative Studies in Society and History*, 3(1), 49–68.

Megoran, N. 2006. For Ethnography in Political Geography: Experiencing and Re-Imagining Ferghana Valley Boundary Closures. *Political Geography*, 25(6), 622–40.

Mountz, A. 2010. *Seeking Asylum: Human Smuggling and Bureaucracy at the Border*. Minneapolis: University of Minnesota Press.

Newman, D. and Paasi, A. 1998. Fences and Neighbours in the Postmodern World: Boundary Narratives in Political Geography. *Progress in Human Geography*, 22(2), 186–207.

Norman, E. 2012. Cultural Politics and Transboundary Resource Governance in the Salish Sea. *Water Alternatives*, 5(1), 138–60.

Paasi, A. 1996. *Territories, Boundaries and Consciousness*. New York: John Wiley and Sons.

Pickering, S. 2006. Border Narratives from Talking Security to Performing Borderlands. *Borders, Mobility and Technologies of Control*, edited by S. Pickering and L. Weber. The Netherlands: Springer, 45–62.

Prokkola, E.K. 2009. Unfixing Borderland Identity: Border Performances and Narratives in the Construction of Self. *Journal of Borderlands Studies*, 24(3), 21–38.

Prokkola, E.K. 2008. Border Narratives at Work: Theatrical Smuggling and the Politics of Commemoration. *Geopolitics*, 13(4), 657–75.

Rajah, A. 1990. Ethnicity, Nationalism, and the Nation-State: The Karen in Burma and Thailand. *Ethnic Groups across National Boundaries in Mainland Southeast Asia*, edited by G. Wijeyewardene. Singapore: Institute of Southeast Asian Studies, 102–33.

Rumford, C. 2008. Special Issue: Citizens and Borderwork in Contemporary Europe. *Space and Polity*, 12(1), 1–12.

Salter, M.B. 2012. Theory of the /: The Suture and Critical Border Studies. *Geopolitics*, 17(4), 734–55.

Scott, J.C. 1998. *Seeing Like a State: How Certain Schemes to Improve the Human Condition Have Failed*. New Haven: Yale University Press.

Scott, J.C. 2009. *The Art of Not Being Governed: An Anarchist History of Upland Southeast Asia*. New Haven: Yale University Press.

Sturgeon, J. 2004. Border Practices, Boundaries, and the Control of Resource Access: A Case from China, Thailand and Burma. *Development and Change*, 35(3), 463–84.

Varis, O., Tortajada, C., and Biswas, A.K. (eds). 2008. *Management of Transboundary Rivers and Lakes*. Berlin: Springer.

Vandergeest, P. 2003. Racialization and Citizenship in Thai Forest Politics. *Society & Natural Resources*, 16(1), 19–37.

Vandergeest, P. and Peluso, N.L. 1995. Territorialization and State Power in Thailand. *Theory and Society*, 24, 385–426.

Vandergeest, P., Bose, P., and Idahosa, P. 2007. *Development's Displacements: Ecologies, Economies, and Cultures at Risk*. Vancouver: UBC Press.

Walker, A. 1999. *The Legend of the Golden Boat*. University of Hawaii Press.

Wong, C.M., Williams, C.E., Pittock, J., Collier, U., and Schelle, P. 2007. World's Top 10 Rivers at Risk. *WWF*. Available at: http://awsassets.panda.org/downloads/worldstop10riversatriskfinalmarch13_1.pdf [accessed: August 2013].

Wijeyewardene, G. 1991/2002. The Frontiers of Thailand. *National Identity and its Defenders: Thailand Today*, edited by C. Reynolds. Chiang Mai, Thailand: Silkworm Press, 126–54.

Wilson, C.M. and Hanks, L.M. 1985. Burma–Thailand Frontier Over Sixteen Decades: Three Descriptive Documents. *Research in International Studies Southeast Asia Series*. Ohio: Ohio University Press.

Winichakul, T. 1994. *Siam Mapped: a History of the Geobody of a Nation*. Hawaii: University of Hawaii Press.

Winichakul, T. 2000. The Quest for "Siwilai": A Geographical Discourse of Civilizational Thinking in the Late Nineteenth and Early Twentieth-Century Siam. *The Journal of Asian Studies*, 59(3), 528–49.

Chapter 7

Crossing Lines, Crossed by Lines: Everyday Practices and Local Border Traffic in Schengen Regulated Borderlands

Judith Miggelbrink

Prologue

> *'Have you been ill this winter? No? Do you have many absences?'*
> *'Have what?'*
> *'Absences?'*
> *'I can't hear you.'*
> *'Have you absences from school?'*
> *[...]*
> *'What's new, Tolica?'*
> *'What?'*
> *'What's new? Nothing?'*
> *'The godmother is at home now.'*

Introduction

The preceding conversation stems from the documentary "podul peste tisa/The Bridge" directed and produced by Ileana Stănculescu (Romania, 2004). A woman was standing on the southern—Romanian—bank of the river Tisza (called Tisa in Romanian and Tyssa in Ukrainian). From there, she shouted across the rushing waters to her son, Tolica. Standing on the northern—Ukrainian—bank of the river, Tolica struggled to understand his mother. Their conversation, however trivial and commonplace, took place in 2003/04 when Ileana Stănculescu produced the documentary. By then, a new bridge already spanned the Tisza between the Romanian town of Sighetu Marmatiei and Solotvyno in Ukraine. Since 2002, the bridge connected the areas on both sides of the Tisza that often are described as part of the same cultural and historical region: the land of Maramureş (Ţara Maramureşului). Both parts were physically separated during World War II when almost all bridges were destroyed. The physical separation was followed by a political one: Whereas the northern part was incorporated into the Soviet Union, the southern parts stayed with Romania. People who lived in this area for generations were separated and interactions between them were cut, although many of them had family ties on the other side and wished to maintain their relations. However, after 10 years of negotiation, planning, and finally building the bridge, "maintaining" relations across the river still meant shouting over the waters. Why?

Following the completion of the bridge in 2002, it took another five years until it was officially opened as a border crossing point. Until 2007 people from both countries were still kept at a distance. Due to relatively liberal border management at that time, by opening the border crossing point the bridge and its surroundings immediately turned into a site of trading, shopping, and commuting. However, the bridge has been closed and opened every now and then for reasons that have never become entirely transparent (for a short summary of the rather complex history of the contemporary bridge, see Wust and Zichner 2010). Ironically, at the same time, Romania transformed the border to Ukraine according to the strict and homogenous border regime that the Commission of the EU has applied from the "waters of the Mediterranean to beyond the Arctic circle."[1] Romania—as well as Bulgaria—has implemented the Schengen rules, but still may only issue national visas instead of Schengen visas to technical procedures that do not meet EU standards so far.

This chapter illustrates how state borders in Central and Eastern Europe that were transformed according to the standards of the European Schengen border regime function within the everyday lives of people. My focus is on two overlapping aspects: On the one hand, it is on the local border traffic (LBT) regime that has been developed by the EU Commission to enable and re-organize cross-border activities of people living in the borderlands alongside the EU borders. On the other hand, it is on small-scale trade as a concrete appropriation of the border and border-dependent practice that has made the border a transitory space offering income opportunities for individuals and households under precarious economic conditions.

In order to elaborate on the role of borders and bordering in everyday life, I draw on governmentality to conceptualize territory and the border in structuring social/society relations. In the following paragraph, I briefly depict a macro perspective of border development in Central and Eastern Europe to demonstrate changing conditions of border crossings that were implemented at the Eastern borders of the Baltic States, Poland, Slovakia, and Hungary. Having set the stage, I then use a micro perspective to analyze some of the effects the changing border regime has on everyday practices of people in the borderlands. In doing so, I focus on a specific group of people: small-scale traders and entrepreneurs living in close proximity to the border and their economic practices that are extremely sensitive to even the modest changes of border regimes.

In this chapter, I draw mainly on fieldwork that was carried out in 2008–2009 when these changes were highly present and widely discussed amongst the

---

1   Homepage of the European Union, Justice and Home Affairs: http://ec.europa.eu/justice_home/fsj/freetravel/ rights/fsj_freetravel_rights_en.htm. Accessed: May 25, 2010.

border population that was confronted with what some of them called "the new Berlin wall." Informal cross-border trade of consumer and household goods as well as food and other produce is a widespread phenomenon at these borders that is well researched in geography, social anthropology, social science, and Central and Eastern European studies (for example Thuen 1999, Hapke 2001, Hohnen 2003, Holtom 2004, Round et al. 2008, Polese 2012, Weiss 2012, Bruns et al. 2013). Though my material was mainly generated in the specific field of small-scale economic practices and entrepreneurship across borders, my focus is not on practices of small-scale trading and smuggling as such (for a more specific analysis, see Bruns et al. 2013) but on how cross-border mobility and interaction are structured and attempts of structuring are dealt with. The European external border is a highly informative example of refining control and surveillance as local border traffic belongs to the "issues that have become rather problematic in the post enlargement era" (Papagianni 2006: 143)—an era that is characterized by a deepening of the gap between those living inside the EU and outsiders and by a strict, visa-based border regulation thus ringing down "a 'paper curtain' of insurmountable visa requirements" (O'Connell 2008: 119). In the last subsection, I discuss the structuring capacity EU local border traffic regime in the making exerts on people in terms of its underlying juridical and disciplinary logics.

The regional settings of my chapter are along the external border of European border (see Figure 7.1). Particularly, I draw on material from interviews and group discussions at the Polish–Belarusian border, the Polish–Ukrainian border and the Romanian–Ukrainian border generated through a comparative research project on cross-border small-scale trade and entrepreneurship (see Bruns, Miggelbrink and Müller 2011; Müller and Miggelbrink 2013, in press).[2] Interviews and group discussions took place in small and medium-sized towns close to the border: in Białystok and Przemysl (Poland), in Sighetu Marmatiei (Romania), in Schowkwa and Solotvyno (Ukraine) as well as in Brest (Belarus).

In order to approach the everyday and its entanglement with the Schengen border, in the following subsection I apply a theoretical approach on borders as a spatial means of governing, i.e. as a power technology and relate it to practice theory.

Border/ing as Power Technology: Conceptual Remarks

For some time now, authors have emphasized that borders are not given entities of social life. Instead, they regard them as both a means and outcome of processes of relating power, space, and people. In order to overcome a static notion of *border*, they shifted their perspective towards a procedural understanding of bordering (Van Houtum and Van Naerssen 2002, Van Houtum and Pijpers 2007, Vaughan-Williams 2008, Van der Velde and Van Naerssen 2011). While these scholars

---

2 Originally, the project also included the Finnish–Russian border which is not discussed here. For an extended analysis see Müller 2013.

Figure 7.1   The Schengen area

mainly worked on state borders and the European border, respectively, the notion of bordering is not limited to the sphere of practices of statehood. On the contrary, bordering processes have been addressed as a basic operation of distinguishing in social life, that *inter alia* becomes manifest in a state's bordering policies but is not exclusively bound to state politics. As a consequence, scholars became more attentive to the how of bordering instead of the where and what (see Van der Velde and Van Naerssen 2011: 220). In particular, it is the how that requires further explication.

I approach the question of how bordering takes place from a power-technology perspective based on Foucault's understanding of governing social relations (see Belina and Miggelbrink 2013: 125–9). From his work, especially from his lectures on biopolitics (1977–78; Foucault 2007), I adopt his framing of power. In his lecture, he defines power not as an asset or a capacity of someone or something but as constantly exerted through pervading, enabling, and forming actions and identities as well as individuals and populations. His attempt to fathom the nexus of power and space is of special interest with regard to borders and processes of bordering. According to Foucault, space is crucial to how power is exercised and how power is involved in governing of the self and others, or as he puts it, "problems of space are equally common to all [logics of power, JM]" (Foucault 2007: 26). Spaces, therefore, are neither stages nor just products of social action but power techniques through which social relations are governed. He identified three logics of power: the juridical mechanism differentiating the permitted and the prohibited, the disciplinary mechanism that aims at transforming the individual (the culprit, the pupil) according to a given norm(ality), and the security mechanism striving for control over (future) events. All three modes of governing action, individuals, and events lean on certain spatial forms: the juridical mechanism on the geographical scope of the law, which is the borders of a territory, the disciplinary mechanism on structured spaces of architecture, and the security logic of power on milieus.

As a means of a juridical mechanism of power, borders serve to establish, demarcate, and maintain a territory. Therefore, they are both medium and effect of territorializing endeavor defined as "the attempt by an individual or a group to influence, affect, or control objects, people, and relationships by delimiting and asserting control over a geographic area" (Sack 1983: 56). Though rather broad in scope, Sack's definition underlines that territorializing attempts are always driven by actors and interests. The "geographic area" through which individuals and groups are governed is a limited space apportioned by a border that serves as a spatial means of territorialized control and regulation. Bordering, thus, is a spatial, but rather unspecific, instrument that can be exploited for a variety of purposes. While Sack's definition of territorializing leaves its function open, writers in the field of political geography and critical geopolitics examining the state-space nexus have addressed it more specifically. Territorializing and bordering necessarily underpin a state's striving for maintaining control and regulation, its striving for maintaining sovereignty, or, as Painter (2010: 1095) put it, "(t)he spatial extent of state sovereignty is coterminous with territory." A state's border marks the spatial scope of the law and, thus, indicates a "radical rupture in the nature and intensity of power."

Beyond its function as markers and maintainers of sovereignty, state borders—as well as borders in general—should also be considered as a spatial means of governing people's identity, or, in Foucauldian terminology, as disciplinary mechanisms of power. As such, borders, including state borders, instate categories of identity and belonging with far reaching consequences concerning social orders of inclusion and exclusion.

Finally, borders also serve as a means of what Foucault has called the dispositive of security: affecting conditions of societal life in way that keeps things running. This third mechanism of power relates to calculation and techniques of calculation; that is, counting, mapping, working on statistics, etc. The security logic is a way of conceptualizing events, things, and persons by putting them "within a series of probable events. [...] the reactions of power to this phenomenon are inserted in a calculation of cost. Finally [...] one establishes an average considered as optimal on the one hand, and, on the other, a bandwidth of the acceptable that must not be exceeded" (Foucault 2007: 20–21). The spatial form that organizes the "optimum" is the milieu which is not a specific space but a relational combination of natural and artificial givens towards certain ends. Borders in the light of the dispositive of security may serve as a means for maintaining the acceptable (e.g. in the field of migration).

Concerning state borders—and borders of supra-state entities—bordering is closely related to territorializing processes. Nevertheless, while state bordering is central for spatially demarcating the realm of territorialized power, it has effects that go beyond maintaining of sovereignty. State bordering entails processes of individualization in that it regulates state-related identities. And state bordering and territorializing is both an outcome and a medium of how processes of production and reproduction, economic and social circulation are organized and regulated. This has lately been underscored by Elden, who pleads for an understanding of territory as power technology (Elden 2010). Taking up Foucault's concept but also going beyond it, Elden identifies the territory as "a rendering of the emergent concept of 'space' as a political category: owned, distributed, mapped, calculated, bordered and controlled" (Elden 2010: 810) or as a strategy of governing insofar as it enables regulation. Based on this, I understand bordering as a spatial (and calculative) means of safeguarding and maintaining processes of circulation in what the EU defines as its inner sphere. Though Harvey, for example, criticizes Foucault's understanding of space for privileging a Newtonian concept of absolute space and thus neglecting the socially produced nature of it (see e.g. Harvey 2007), it offers a useful approach to understand the geopolitical and biopolitical dimension of state borders in general as well as of the transnational border of the European Union (see Walters 2002, Coleman and Stuesse, this volume). For this chapter, I focus on the disciplinary effects of bordering processes that are realized through politics of identity and social categorization. With regard to the European Union, Van Houtum and Pijpers (2007: 297) stated that "the current spatial imaginative bordering process [...] rests upon the colonization of friends as members or associated members (Bauman 1990), among whom common assets of knowledge and wealth are constructed and distributed. To the Other, residential rights are granted only if such an extension of rights does not threaten the existing order (Bauman 1990). From this perspective, bordering is a permanent effort to implement and stabilize a distinction between those who are regarded as threatening and those who are not. Likewise, with regard to the European Union, Balibar and other authors (Paasi 1998, Newman 2006) underline that border politics indeed has established rather clear-cut categories of identity and belonging based on "citizenship." Since the Treaty of Maastricht (1992), when the EU introduced the new category of the EU citizenship—which is the first transnational citizenship worldwide (see Buckel and Wissel 2009)—an additional layer was put on top of the national citizenship. This has led to a growing gap between those who are included by the border (the citizens of the EU) and those who are excluded (those who do not enjoy this status) (Balibar 2002). Consequently, categories of identification have been trans-nationalized and scaled-up. Moreover, those who have a non-EU citizenship are divided into citizens of visa-requiring and visa-exempt states. Whereas visas are waived for the citizens of the Western Balkan states, for instance, citizen of eastern European states require a visa. The excluded individuals, thus, are structured according to stratified rights (Buckel and Wissel 2009) that mainly mirror the geopolitical image of the state of origin (Bekus-Goncharova 2008). The identity of the Other, however—the "identity of strangers," as van Houtum and Pijpers put it, (2007: 297–8) "is not their choice." Politics of identity with regard to the border of the European Union means both that a social order is established by categorization and that the capacity for influencing these categorizations is rather limited for the Other.

In addition to an understanding of border/ing as a power technology, I refer to practice theory as another conceptual source because I do not want to treat empirical example as mere illustration of theoretical abstraction(s) but want to fully acknowledge the role of practice in structuring the social word. As Jones and Murphy (2011: 367) state for economic-geographical phenomena, "[…] the study of practice can complement existing explanations for economic-geographical phenomena (e.g. institutional, radical, or relational) by providing an analytical 'object' whose study can demonstrate how higher-order phenomena such as institutions, networks, class structures, and gender inequalities are enacted, reproduced, and/or transformed through the everyday actions embedded within them."

What a border is in the end and how it works in concrete situations is neither entirely determined by geopolitical and biopolitical will and reason, nor can it be reduced to the juridical regulations of the border, its disciplinary attempt, or its role as an element of building milieus (and their legitimizations). Furthermore, it is not entirely determined by the capacities of technologies applied to anticipate migration flows, to identify, to filter, and to sort out people (Bigo 2005). The border is realized in the very moment of border crossing when state sovereignty and everyday practices meet and the border becomes a medium *"through which* institutions, associations or legal machineries operate" (Nadel 1956: 171–2, quoted by Donnan and Wilson 2003: 11, emphasis in original).

Entanglements of the Everyday and State Sovereignty: The Contested Space of the Borderland

Areas close to the border are often marginal and peripheral parts of a state's territory in terms of socioeconomic development as a location on the edges of the state often comes along with a comparatively low economic performance. However, there might also be economic advantages including cross-border consumer tourism, which is a side effect of structural differences between neighboring states. At the same time (and partly because of their low economic performance) borderlands have increasingly been regarded as laboratories of new forms of cross-border governance and cooperation (e.g. Euregio); in this respect, they are often treated as areas of expectation. Here, everyday cross-border contacts are given a strategic position: Good neighborhood relations should not only be built on cooperation between local authorities but also be built from below, from contacts between people in the fields of culture, education, and sports. The borderlands are also generated through practices that challenge the state (e.g. smuggling). And finally, the borderlands are often areas of ethnic mixing inhabited by people potentially perceiving themselves as ethnically distinct or are regarded as such and thus suspected of not being loyal to the state they are living in and to its majority society. Nikiforova reminds us that groups or interests in contradicting and challenging the state often develop "counter-narratives of place, identity and border in an effort to renegotiate its meaning and reappropriate the border for their own purposes" (Nikiforova 2005: 193). Therefore she concludes that "every border is polysemic and contested" (194).

The EU Commission is concerned that there are a broad variety of motives that have led to the intense cross-border movement. The ambivalence of everyday cross-border activities derives from different and contradictory orders of judging them: Which kinds of contacts are welcomed by whom? What is legal? What is accepted as legitimate? Even smuggling—a cross-border practice that challenges the state's authority to regulate cross border commerce—might somehow be accepted in precarious times, under certain conditions and by certain actors (examples are given by Bruns 2010).

In short, borderlands are not fixed or coherent territories but rather the effects of dynamic and overlapping socio-spatial practices, expectations, perceptions and decisions based on, referring to and resulting from a state's bordering politics. Borderlands could not be thought of without a sovereign entity establishing and maintaining a border. Nevertheless, though borderlands are inevitably based on state practices they are not entirely defined by the state as they result from practices of appropriating the border. Taking into account everyday practices of appropriation, borderlands are related to different scales: They are formed by tensions between a state's ongoing attempt of securing its sovereignty on the one hand and the nested necessities, claims, demands and desires of the everyday that imbue the local "where we live our daily lives" (Taylor 1982) on the other.

In the following two main subsections of my chapter, I investigate the realization of the border from two perspectives: First, I choose a macro-perspective to recapitulate some main steps of the development of a common border regime and to position local border traffic within the EU's Geo-economic-political project. Second, I turn to the micro-perspective of small-scale traders to illustrate how they have dealt with the border in their everyday practices. In the aftermath of the political changes in Central and Eastern Europe in 1989–91, local border traffic had gained new momentum as several borders strictly closed before eventually had become lines of contact and exchange. Daily life started or continued to cross the lines, taking advantage of "resourcing borderlands" (Weiss 2012: 213). This holds for the Polish–Ukrainian border and the Polish border to the Russian exclave of Kaliningrad where small-scale trade in close vicinity of the border has become a widely used opportunity to overcome precarious economic conditions in the aftermath of socio-political changes (see Haase et al. 2004, Wagner 2010, Bruns 2010). However, new borders also were erected when the Soviet Union collapsed and former Soviet republics became independent states, such as Estonia, Latvia, Lithuania, Belarus, Ukraine, Moldova, and several central Asian states. New nations sought sovereignty, integrity, and identity for which safeguarding borders played a substantial and symbolic role (e.g. Kuus 2007 for the case of Estonia). Occasionally (and partly as a side effect), this led to a separation of places formerly interlinked through people living their daily lives. Throughout Europe there are examples of places where the Schengen border now cuts through regions; one is the region of Maramureş mentioned in the introduction to the chapter. Another one is the region of Setumaa in the area of Voru and Petsori in the Estonian–Russian borderlands (see, for example, Kaiser and Nikiforova 2006). In these regions, the population has preserved a certain notion of shared history and belonging which have been subjected to the formation of the EU and its territorializing and bordering claims. After this period of territorial re-ordering—a period of drawing new lines of separation but also of comparably liberal bilateral border regimes—a new era dawned when in 2004 the external border of the European Union shifted towards the East. To reveal the significance of this process, one has to go back into the history of the entanglement of European integration and border policy.

"From the waters of the Mediterranean to beyond the Arctic circle": Stages of Border Development and its Effects on Borderlands and Local Border Traffic, from a Macro Perspective

Towards Common and Homogenous Border Regulations

When the Schengen Agreement was signed in 1985 in the Luxembourgian border village Schengen, it was not part of the existing treaties of the European

Community but a multilateral agreement of five of the six founding states of the European Community: Belgium, the Netherlands, Luxembourg, France, and Germany (see Table 7.1 for an overview of the main steps and treaties). They took the initiative to provide the necessary prerequisites for a unified (single) market that was assumed to strengthen the EC as a globally competitive geo-economic actor (see Altvater and Mahnkopf 2007). The preconditions—free movement of people, finances, goods, and services—included the future abolition of borders between signatory states. As its counterpart, signatory states agreed to develop a border regime at the borders between their territories and non-signatory states that would meet the security interests of the signing governments. Though the agreement was not formally part of the EC, it was both an instrument to overcome economic stagnancy in member states of the EC and an instrument to consolidate European integration. Details were elaborated in the Schengen Convention signed in 1990 and entered into force in 1995.

In the meantime, the European Community member states founded the European Monetary Union (Treaty of Maastricht, 1992) by which the European Union as a follower of the European Community took shape. Both the Schengen Convention and the Monetary Union paved the way for a unified market. As their outcome "the economically relevant border functions to the external EU border" were relocated (Belina and Miggelbrink 2013: 130). The Treaty of Maastricht together with the following Treaties of Amsterdam and Nice defined and consolidated the so called three pillar model, a model of shared responsibilities in the fields of Justice and Home Affairs, Police and Judicial Co-operation in Criminal Matters as well as Common Foreign and Security Policy. Border functions were allocated to all fields thus often lacking coherence. Nevertheless, by this, states had agreed to partly shift sovereignty towards the transnational scale of the European Union that became in charge of further developing a consolidated regulative basis for the EU, the Schengen *acquis* (the cumulative set of agreements, legislation, and court decision that form EU law). Additionally, by the Treaty of Amsterdam the inner sphere of the Union rhetorically became the "Area of Freedom, Security and Justice" which has been put against a clearly separated outer sphere.

Though the EU Commission has continuously developed a shared corpus of rules and legislation since the 1980s (see Dubowski 2012), the process gained momentum in preparation of the accession of several Central and Eastern European states to the Union. In the meantime the Schengen *acquis* had become part of the general *acquis communautaire*, so that candidate countries were also obliged to implement the border related *acquis*. Until the end of 2007, the new member states except of Romania and Bulgaria had succeeded in fully adjusting all border relevant procedures including their databases to Schengen standards. This led to a new and, at first glance, rather simple looking picture of bordered territories: Schengen member states on the western side of the line; all non-member (so called third) countries on the eastern side. Due to a "one-fits-all" politics, as the EU calls it, all former regulations concerning border passes, migration cards, entry and exit

Table 7.1   Steps of European integration, border policy and local border traffic

European integration since the 1980s	Border policy	Local Border Traffic (LBT)
1980s Preparation of a unified ("single") market	**1985** Schengen Agreement ("Schengen I") on a future abolition of borders between signatory states (France, Belgium, Netherlands, Luxembourg, Germany)	
1992/1993* European Monetary Union (Treaty of Maastricht) introducing the European Union based on a three-pillar structure	**1990/1995** Schengen Convention ("Schengen II") abolition of internal border controls between signatory states and relocation of border functions to external borders	LBT mentioned in art 3 (need for future solutions)
1997/1999 Treaty of the European Union (Treaty of Amsterdam) consolidating the European Union; EC and EU became parallel structures	Incorporation of the Schengen Convention into the acquis communautaire; introducing of the geopolitical semantics of an "Area of Freedom, Security and Justice"	
2001/2003 Treaty of Nice consolidating the EU		
2003 Settling of the European Neighbourhood Policy (ENP)	**2002** Communication on integrated border management; need of facilitating border-crossing for bona fide third-country nationals underlined	**2002** Inventory of existing inter-governmental LBT arrangements
2004 Ten states acceded to the EU (Estonia, Latvia, Lithuania, Poland, Czech Republic, Slovakia, Hungary, Slovenia, Cyprus, Malta)	New neighboring states adjacent to the Union	
	2005 Treaty of Prüm ("Schengen III"), stepping up of cross-border cooperation, particularly in combating terrorism, cross-border crime and illegal migration; foundation of the European border agency FRONTEX	**2006** Regulation(EC) 1931/2006 laid down common rules for LBT

European integration since the 1980s	Border policy	Local Border Traffic (LBT)
2007 Bulgaria and Romania member of the EU	**2007** Territorial shift of the Schengen border towards the east (Romania and Bulgaria still not fully integrated)	
2007/2009 Treaty of Lisboa, abolishing of the previous three-pillar model of the EU; the EC was finally substituted by the EU	EU formally became a foreign affairs actor by introducing a High Representative of the Union for Foreign Affairs and Security Policy; intensified debates on temporary border controls at national borders (suspension of the Schengen regulations)	**2008** First bilateral agreement based on Regulation(EC) 1931/2006 entered into force between Hungary and Ukraine and Slovakia and Ukraine, respectively; further agreements followed or are under negotiation
2013 Croatia member of the EU	**2013** Smart border package announced	

Note: * Signed/entered into force.

stamps, as well as liberal practices of control had to be abolished and replaced by the Schengen border code.

What are the main outcomes of this process so far? Three of the juridical process should be highlighted. First, developing a common border policy means a double shift. There was a scalar shift from the sphere of the national towards a transnational regulation realized through a shift of sovereignty towards the transnational entity of the EU though its status concerning statehood is still open and unclear. This up-scaling of sovereignty that came along with the internal market has turned European integration into a territorial project; its "substantive content" (Cox 2002: 11) being driven by geo-economic interests (Heeg and Ossenbrügge 2002). Tolica and his mother were literally standing on both sides of the emerging territorial project of the EU: the side of the enlarged single market and the outside. Beyond the scalar shift, there has been a series of territorial shifts as the border was relocated several times, thus including new territories into the Area of Freedom, Justice, and Security and producing new neighboring countries and new borderlands.

Second, the process of developing a common border is a process of ongoing securitization (Huysmans 2000, Belina and Miggelbrink 2013). Apap et al. (2001: 1) describe its underlying logic "strongly influenced by widespread fear of uncontrolled immigration from beyond EU territory." However, even though the European border regime has been characterized as an excluding, at best highly selective instrument (Van Houtum and Pijpers 2007: 299) and a "security perimeter strategy" (Brunet-Jailly 2006), the relation to its outside is more complex. On the one hand, discourses of securitization in the name of securing freedom, security, and justice indeed have strongly imbued border policies. On the other hand, building efficient contacts including sustainable cross-border relations to neighboring states has also been highly valued (see Balcsok, Dancs and Koncz 2005, Dimitrovova 2008, Dubowski 2012). Browning and Joenniemi (2008: 531) identify this contradictory relation as the "integration-security dilemma": security interests hampering closer cooperation and cooperation threatening security.

Third, as a consequence of this dilemma, the permeability of the border for (ordinary) people has become more and more subject to management and planning. Which kind of contacts are desired and should therefore be developed by which kind of instrument? How can the exclusion of undesired travelers be guaranteed? To whom should visa facilitation be granted? Which kind of relations should be institutionalized? These questions are not only relevant with regard to the formation of border regulation but are part and parcel of the transforming nature of the EU as a geopolitical actor. As such, the EU not only aims at carefully establishing relations to neighboring states but strives after an "increasingly important role in governing its immediate 'Neighborhood'" (Białasiewicz et al. 2009: 79) to protect the common internal sphere from external "'hard' threats" (ibid.). To ensure a common welfare. Moreover, in this "last round of enlargement," as Browning and Joenniemi (2008: 630) put it, "the EU has been compelled to coin and increasingly explicit geopolitical doctrine to deal with the challenges posed by its new neighbors." Since 2003 this striving for balance was bundled in the European Neighbourhood Policy (ENP) for which a so-called Communication from the EU Commission to the European Council and the European Parliament "Wider Europe—Neighbourhood: A New Framework for Relations with our Eastern and Southern Neighbours" (COM 2003: 104) is central. By this, the EU aims to govern the social multitude outside its own territory and beyond its border. This was addressed as extraterritorial or external governance (see, for example, Lavenex 2006, Rijpma and Cremona 2007, Lavenex and Schimmelfennig 2009). Though the ENP was blamed for its failure to transcend the security-integration dilemma, its "friendly Monroe doctrine" (Emerson 2002) is still the leading programmatic to organize neighborly relations.

To sum up this part, the history of the European border regime can be understood as a process of homogenizing, up-scaling, and securitizing surveillance and control that leads to contradictions concerning the possibility of cross-border relations. In order to transcend the dilemma, the European Neighbourhood Policy (ENP) sets the discursive framework to counterbalance the shortcomings of securitization and to reorganize neighborly relations at different scales. The main purpose of the ENP is to offer and finance programs and projects in different fields (market access, economic integration, and development, mobility of people, and a greater share of the EU financial support in the neighboring countries). The ENP-Instrument explicitly aims at promoting economic and social support in border areas, at ensuring secure borders, and at supporting people-to-people contacts. However, the ENP in turn has produced its own shortcomings as it follows, in the EU jargon, a "communitarized" (and thus centralized) agenda that unilaterally offers a partnership at a distance, without perspective to accession. In the following subsection, I analyze the effects of this process for the borderlands and how the EU reacted to these effects by initializing a local border traffic regime.

Incorporation of Borderlands and Local Border Traffic into the Dilemma of Security and Integration

The risk of breaking cross-border relations by a strict and securitized border regime as well as the need to find a solution in accordance with security interests was identified in Article 3 of the Schengen Convention, which stated that "(m)ore detailed provisions, exceptions and arrangements for local border traffic ... shall be adopted by the Executive Committee."[3] Since then, the EU Commission repeatedly underlined the "need to develop rules on local border traffic in order to consolidate the Community legal framework" on external borders (Regulation (EC) 1931/2006: L405/2). Consequently, the legal framework in the making has subjected local border traffic to the interest of European border policies and, therefore, has appropriated the legal instrument of local border traffic. However, as Table 7.1 shows, there was a time offset between the common external border that was mainly developed during the 1990s and the effort to meet the need of facilitating local border traffic regime which started only about 2003.

A first inventory on existing local border traffic provided an overview of former bilateral agreements (SEC(2002)947, Annex I and II). All the arrangements listed in the document somehow facilitated border crossing for people in the borderlands: either by indicating special border crossing points or by issuing certain documents (for example, special border permits) or by other kinds of privileged treatment. The EU Commission regarded the mosaic of bilateral agreements as a potential threat to security:

> Although some bilateral agreements regulating the issue do exist, there is the need to have a uniform, horizontal and coherent approach, as well as to envisage rules covering different kinds of situations. These would include the need to establish arrangements to cover the situation of visa-obligated third countries. This will be of immediate relevance to the future Member States who will not only need clear guidance as regards rules on local border traffic with their neighbouring countries, but also need to be closely involved in drawing up such rules. (SEC(2002)947: 15)

---

3   The Convention is published as part of the Schengen Acquis in the Official Journal of the European Communities on September 22, 2000.

On the eve of the access of 10 states to the Union, eight of them—Estonia, Latvia, Lithuania, Poland, Poland, Czech Republic, Slovakia, and Hungary—located in Middle Eastern Europe, local cross-border movement became an urgent issue for the European Commission. At that time, the Commission had to solve the problem of cross-border movements resulting from the intense border traffic between states ready to access to the Union and their eastern and southern neighboring states and the economic discrepancy along the future border of the EU. This was challenging insofar as, on the one hand, local solutions bore the risk of issuing valid travel documents below Schengen standards which would allow holders to travel throughout the common territory. This was not acceptable from the perspective of the Commission. On the other hand, the Commission had to find solutions for intergovernmental cooperation and for contact and cooperation on regional and local scales, including trade and labor markets.

> [The] issue assumes a particular importance in the perspective of enlargement, since cross-border movements between candidate countries as well as between candidate countries, on the one side, and their neighbours, on the other side, are very important in number. Efficient rules for 'local border traffic' will promote economic development of border regions and serve as an instrument to reduce gaps in economic standards. Moreover, transfrontier workers are often a needed element in the border regions of the Member States. (SEC(2002)947: 3)

A report on the development of a so called integrated border management that was published almost simultaneously again underlined the necessity to harmonize local border traffic, that is to "(i)dentify principles and adopt common measures on 'local border traffic,' particularly with a view to enlargement. The Commission intended an initiative aiming to better define the fundamental principles and procedures of such a system and, if necessary, to prepare for agreements between the Community and neighboring third countries" (COM(2002)233: 11). As a consequence, the following communication set a discursive framing of future relations to the then neighboring areas that also included a perspective for local border traffic regime. It emphasizes the role of cross-border contacts and movements for "any regional development policy" (COM(2003)104: 11). Therefore, "bona fide third-country nationals living in the border areas that have legitimate and valid grounds for regularly crossing the border and do not pose any security threat" should enjoy facilitation (ibid.) that would both circumvent visa application as well as expedite handling at the border.

This resulted in the 2006 Regulation 1931/2006 that provides the legal framework for further bilateral agreements on local border traffic. Hereby, the European Community finally put member states in the position to negotiate new bilateral agreements with non-member states. Any existing agreement, like between Slovenia and Croatia, had to be amended according to the standards defined by Regulation (EC)1931/2006. Conditions that have to be met by people who want to take part in local border traffic are: 1) They have to be a resident of a border area to be defined by the states in accordance with the regulation, 2) they have to prove eligible grounds to cross the border on a regular basis, 3) they must not be under Schengen Information System (SIS)[4] alerts or bans, and 4) they must not be "any threat to the public policy, public security, or public health of any Member State" (Peers 2011: 211). By this, the process of transforming local border traffic into a "derogation from the general rules" (Dubowski 2012: 371) was completed. As the LBT is subsequent to highly securitized migration control interests, the exemption itself follows "standardized rules" (Peers, Guild and Tomkins 2012: 206f) protecting it from unauthorized use. Figure 7.1 shows the bilateral agreements signed so far.

The Community's efforts on developing a local border traffic regime have been interpreted as a step towards overcoming "a 'paper curtain' of insurmountable visa requirements" (O'Connell 2008: 119). This interpretation is in line with the Commission's own description of the objectives related to a local facilitation of the Schengen Border Code: "to ensure that the borders with its neighbours are not a barrier to trade, social and cultural interchange or regional cooperation" (Regulation (EC)1931/2006: L405/2). This was underlined by O'Connell (2008: 120) who observed that "some member states are more inclined than others to pursue the implementation of such a regime especially where access to former territories, ethnic diasporas or seasonal labour may be at stake." Whereas this comment understands the LBT regime as serving interests of interaction between citizens of different states, Wesseling and Boniface emphasize the organizational and long-term dimension. For them, "(v)isa facilitation and local border-traffic agreements are short term responses to immediate challenges. They are meant to alleviate the pressure on EU consulates and allow them to deliver their services more efficiently" (2008: 139). Especially with regard to the building of relations to neighboring countries and people, they judge LBT regimes as an interim arrangement. Referring to EU diplomats, they assume that "those types of agreement will lead to a decrease in the workload of consulates since they facilitate the granting of multiple-entry visa. Yet considering the broader ENP framework, those types of agreement do not constitute a long-standing solution for the future of relations with the role ENP region" (ibid.). Sustainable relation would require a complete waiving of visa that should be stay "the ultimate objective" (ibid.).

Appropriating the Border: Everyday Small-scale Trade in the Borderlands

In this subsection, I discuss some of the effects the changing border regime has had on the lives of people in the borderlands of Poland and the Ukraine, Poland and Belarus, as well as Romania and Ukraine. My focus is on small-scale trade

---

4   The Schengen Information System (SIS) is a data base listing people and pieces of goods of interest; the common data base is fed by governmental databases thus mirroring the states' different practices of data collection.

which is, of course, only one way of appropriating a border, and, a very specific one at that. The field study took place in 2008 and 2009 exactly in the time gap described above: the external border had already shifted towards the east. Though the EU Commission had adopted the 2006 Regulation which allowed governments to negotiate facilitations for people in the borderlands, only a few governments had started the process of implementing it. Though Hungary and Slovakia signed bilateral agreements with the Ukrainian government in 2008, negotiations between Poland and Ukraine as well as between Poland and Belarus progressed slowly. I assume that several reasons are relevant here rooted in historical and political relations between Poland and Ukraine. Presumably, it is also of relevance that Ukraine tried to include the metropolitan region of Lviv into the privileged border area going far beyond the 50 km zone. The Belarusian government that signed an agreement with Latvia in 2012 has been hesitant to the EU Commission's offer of partnership based on ENP in general.

The Romanian and Ukrainian government could not yet start negotiations due to the specific status of Romania. Romanian consulates are not allowed to issue the Schengen visa as the Romanian border control system is not yet fully in line with Schengen standards. Nevertheless, Ukrainians applying for a national Romanian visa have to meet the same conditions as for a Schengen visa (in terms of the application process, fees, and visiting the consulate in person). The interviews were conducted during a period of discontinuity and interruption for small-scale traders' economic practices and perception of the border. Before I go more into detail, it is appropriate to define small-scale trade.

Cross-border small-scale trade is not a clearly defined activity but mainly understood as an arbitrage trade carried by people trading goods from one side of the border to the other (Hapke 2001, Hohnen 2003, Holtom, 2004). The amount of goods is often limited to what the term "suitcase trade," another term for it, would suggest. The amount of capital traders have at their command is also limited. The trade itself is widely regarded as risk taking because is comprises as least informal if not illegal elements. The (in)formality of small-scale trade cannot be determined absolutely because "(t)o what extent a certain economic cross border practice is formal or informal depends on the degree of its correspondence with regulations of cross border economic interactions" (Bruns, Miggelbrink, and Müller 2011: 667). That means "trade" and "smuggling" highlight different aspects of a continuum of activities: "Traditionally 'trade' is the legal and 'smuggling' is the illegal means of moving items from one side of the border to the other" (Thuen 1999: 741). Often there exists a distinction between the licit and the illicit as "official rules, structures, and discourses do posit a sharp distinction between law and crime" (Van Schendel and Abraham 2005: 7). Nevertheless, this distinction should not be taken as a given but, instead, be regarded as an outcome of "ongoing struggles over legitimacy" (ibid.). In our fieldwork, we did not start with an a priori distinction but left it open to the actors to characterize cross-border activities.

In our group discussions and interviews we talked to male and female traders. At markets and at border crossing points we got the impression of a female predominance as traders and also observed many elderly (retired) people who were involved in trading activities. Most of the people we met were part-time traders: Either they traded goods only occasionally (for example, when an additional income is needed) or on a regular basis in their free time. The female traders we talked with in Brest (Belarus) were full-time traders; they bought produce on Polish markets and sold it at a market in Brest (Belarus) until the territorial shift of the Schengen border forced them to buy their goods in markets around Moscow.

One of our main results is the variety of tactics and strategies small-scale traders deployed to adjust to local and changing conditions (Müller and Miggelbrink, 2014). Most of the tactics were developed to overcome the obstacles of the borders: techniques of hiding the goods, false declaration, bribing, and tacit knowledge about the border personnel and their shifts. More sophisticated (and strategic) ways of overcoming the border are letters of recommendation that some people carry with them for a variety of situations. Another network-based strategy is to divide a certain amount of goods (shoes, sweaters, and cell phones, for example) into smaller, eligible amounts brought through customs clearance by subcontractors. Tactics and strategies do not react to a border regime in general but are based on local experiences, knowledge and contacts that allow successfully dealing with "arbitrary contextual conditions surrounding the realization of the border regime" (Bruns et al. 2013: 103). Other actors, however, have to adapt themselves in their income securing activities to given structures and conditions. They only have the freedom to exploit the characteristics and imponderables of the border regime tactically, for example by choosing another crossing point or trying to avoid certain customs shifts. In doing so, they can draw advantage for their cross-border activities from favorable situations. However, many actors have to adjust to indeterminate situations and can hardly influence the external conditions at all. It was quite often mentioned by people that they feel delivered up into the hands of an entity completely detached from their own sphere that would never care about the "little man."

Although a "soft border" could make it easier for traders to gain an income compared to those who are struggling to trade across a "hard border" (see Xheneti et al. 2013), a "hard border" is not per se a hindrance to a successful trade. From our fieldwork, there is empirical evidence that stability of conditions is most important to the traders because it allows them to reliably calculate risk and gains. It is not the "hardening of the eastern borders" (Xheneti et al. 2013: 319) in terms of its securitization that per se has had an effect on economic activities but the time and money consuming procedures coming along with. It simply requires a new calculation: Is it (still) worth to invest for a visa and to take the risk at the borders? A Ukrainian entrepreneur from Solotvina explained it as follows:

> Basarab: There are lots of facts. In order to get to the next places from here in Romania you need a visa, you need is an invitation, an invitation costs money. The visa itself costs 55–105 Dollars. A simple visa—if you want it within a week—costs 55 Dollars,

Moderator: mmm

Basarab: fast-tracked, 105 Dollars. The invitation itself, about 20–30 Dollars. Two trips to Černovcy, putting your name down on the waiting list, you have to wait one-and-a-half months for your turn. So, you can see for yourself how you can get to Europe.

When a visa is needed which is expensive and limited to one entry, there is literally no basis for trading.

Economic practices including informal economic activities "are situated in specific and often dispersed geographies—economic, but also social, cultural, and political." Thus economic practices are conceptualized as "a wide range of mechanisms by which individuals and the social units of which they are a part create livelihoods" (Smith and Stenning 2006: 192). As small-scale trade is extremely sensitive to contextual conditions, traders talked a lot about the newly generated exclusive spheres. The territorial distinction and categorization introduced by the Schengen border regime is a major concern of the small traders we interviewed along the current existing or emerging Schengen border. They describe it as a new distinction that does not correspond to the "real" or "proper" spatial order, which they define in terms of cultural, national, or ethnical belonging. Anton, a Ukrainian interviewee, explained: "Nevertheless for us [...] that is, our Romanian places and the places on the other side, and the towns are almost twin towns. In principle we are one nationality, one nation, one culture" (from a group discussion with small-scale traders in Solotvina (Ukraine), July 28, 2008). One spatial order is contrasted with a different, alternative spatial-territorial order, which is taken as proof that the new order is inappropriate. From the point of view of this Ukrainian, the false territorialization could be fixed by a bilateral agreement between the Ukraine and Romania:

> [...] they invite us over. They visit us. In this respect of course ... we are happy that the border was opened. But we also naturally wish—that a zone could be created for us, like for Hungary and for the Poles, at least, at least that this fifty kilometer zone would be introduced for us so the local population ... can travel without a visa.

The true regional order of social life traders were talking about is, of course, not only a place of identity and belonging but also a site of daily activities where economic practices including informal cross-border economic activities could take place.

Discussion and Conclusion: A New Local Border Traffic Regime in the Making? The Political Production of Borderlands According to Schengen

Small-scale trade is part of the social life in borderlands along the external border of the EU. It has been an important though additional source of income for many households and thus been regarded as a means to secure social peace. Therefore, it is not surprising that people actively involved in informal activities did not welcome the Schengen regime. The traders we talked with feel excluded from the European Union and a majority of them would welcome a regional solution of border control based on the EU Regulation of LBT to overcome the impasse of the border. From the perspective of the EU Commission, the new border regime does not essentially aim at denying access to the "area of freedom, security and justice" (Consolidated Version of the Treaty of European Union (also "Treaty of Amsterdam"), C 321 E/10, preamble). Instead, it aims at getting full control over people from the outside, their movements and motives. The procedure of sorting "flows of people" (Bigo 2005: 52) and filtering them according to anticipated motives of movement is a reiterating management task of refining and adjusting the filtering capacity of the border. From this perspective, the local border traffic regime is a new round in the process of refining the border and making it smart according to the objectives of EU policies. Though presented as facilitating cross-border movements of people living in the borderlands, it has a far reaching governmental dimension.

What does this mean for everyday practices in borderlands? The intended local border traffic regime which has to be understood as part of an emerging smart border system is clearly constructed as an "exemption from the general rules for external border crossings" (COM(2011)47: 6), or, as it was described in the 2006 regulation, it "constitutes a derogation from the general rules governing the border control of persons" (Regulation (EC)1931/2006: L405/2). It is an instrument of border facilitation which is under control of the EU Commission—though it is unclear if and how insufficient agreements can be acted on from a supranational European entity (see, for example, Ziltener 2000).

The LBT regime sorts people and potential border crossings by classifying them as desired or to be refused. Nevertheless, first of all that does not mean to differentiate between individuals but to superimpose categories on them by which certain travelers are privileged and certain traveling motives are stream-lined. Although still a regime in the making, a new relation of power and space with regard to borderlands is emerging: The regime is essentially a territorial means of structuring (and probably intensifying) cross-border movements since the eligible population is territorially defined as residents of demarcated administrative areas. In terms of theorizing from a Foucauldian perspective, in a spatially defined eligibility one can see the juridical mechanism of power at work. Certain territories and their residential populations are included in bilateral agreements, others are not. However, spatial eligibility is only a first criterion as people have to apply individually for LBT permit. In that way the LBT regime unfolds a structuring capacity towards the individual based on a disciplinary mechanism of power. Remarkably, it is not regarded as sufficient not to be under SIS alert or bans but people have to disclose their motives for being granted with exemption. Taking part in local border traffic does not depend (exclusively) on personal integrity according to Schengen—meaning that the applicant is not listed on databases feeding the Schengen Information System—but also on one's motives that have to be in line with the expectation of the issuing authority. First reports to the Commission (COM(2009)383, COM(2011)47) stated that in several cases permission was denied because of registered SIS alerts and bans. Although this is still speculative, there are hints that having the wrong or insufficient motives could provoke rejection: The 2011 report of the EU Commission states—without any further comment or explanation—that Slovakia for example has refused issuing of permits because "applicants did not give any well-founded economic reason for frequent border crossing" in 169 cases (COM(2011)47: 4). Romania refused more than 900 applicants "mainly because applicants did not give any well-founded reason for frequent border crossings or were considered to present risks related to irregular migration" (COM(2011)47: 4–5). These are clear hints that facilitation is not granted to residents of the border region if they are not in conflict with the law but only if they claim acceptable motives for crossing the border. In turn, it can be assumed that people will find a way to bring their motives in line with what is expected by the authorities. The disciplinary mechanism at work unfolds on the individual who has to bring his or her cross-border movements in line with the requirements: It is not simply everyone within the area of eligibility who enjoys being visa-exempt since individuals could be excluded from facilitation if they do not fit into a certain scheme of expectation (or simply do not sufficiently adjust to it). It is a scheme that—depending on conditions and contexts—could turn out to be transparent and fair but is also arbitrary and unfair.

Borderlands are zones of sovereignty and subversion. The snapshot from the documentary shows something very banal: A mother and son shouting across the waters who can hardly understand each other. There seems to be no chance for closer (and calmer) personal interaction in their immediate reach. We do not know whether Tolica's mother is involved in small-scale trade (though this is likely given the spatial and temporal context the documentary captured); what we see is a border cutting through their lives. We do not know whether the mother is involved in illicit activities or if the family is used to subvert the border. What we can observe is a hampered and disrupted interaction that is rooted in the specific boundedness of the place that is the bordered land, or, the borderland. In this chapter, I understand borderlands as a spatial form, which is, on the one hand, defined by its proximity to a (state) border and, thus, of special concern with regard to surveillance and control by state authorities, and, on the other hand, not clearly defined in terms of its territorial extension but generated through a broad variety of interests and practices related to the border. I have sought to analyze how the shifted transnational border regime of the EU affected everyday cross-border practices and, moreover, how changes have imbued the way(s) people in the adjacent borderlands felt as the Other. My argument was based on the assumption that border regimes, though never fully determining cross-border interaction, still inheres a far reaching capacity to structure social relations.

From a perspective of the state—and from the perspective of the state-like authorities of the EU—the borderland is a highly sensitive area as it is a crucial site of maintaining state sovereignty. Not only is the border itself subject to particular political and administrative concern but also the territory close to the border where the state literally comes to its end. The local border traffic regime is a spatial means gaining control over movements that aims at balancing the contradiction between securitization and cooperation. As such, it unfolds its own structural selectivity towards cross-border movements of people in the vicinity of the border. How people will deal with these new juridical and disciplinary attempts to structure their doings and sayings has to be proven by future empirical work.

References

Sources from EC/EU Authorities

COM(2002)233 final. Commission of the European Communities. 2002. Communication from the Commission to the Council and the European Parliament towards integrated management of the external borders of the member states of the European Union. COM(2002)233. Brussels. [Online]. Available at: http://eur-lex.europa.eu/LexUriServ/LexUriServ.do? uri=COM:2002:0233:FIN:EN:PDF [accessed: May 30, 2013].

SEC(2002)947. Kommision der Europäischen Gemeinschaften. 2002. Arbeitsdokument der Kommisionsdienststellen. Weiterentwicklung des Besitzstand im Bereich des "kleinen Grenzverkehrs." Brüssel. [Online]. Available at: http://edz.bib. uni-mannheim.de/www-edz/pdf/sek/2002/sek-2002–0947.pdf [accessed: July 22, 2013].

COM(2003)104 final. Commission of the European Communities. 2003. Communication from the Commission to the Council and the European Parliament. "Wider Europe—Neighbourhood: A New Framework for Relations with our Eastern and Southern Neighbours." Brussels. [Online]. Available at: http://eur-lex.europa.eu/LexUriServ/LexUriServ.do?uri=COM: 2003:0104:FIN:EN:PDF [accessed: July 22, 2013].

COM(2003)502 final. Proposal for a Council Regulation on the establishment of a regime of local border traffic at the external land borders of the Member States. 2003/0193 (CNS), 2003/0194 (CNS). Brussels. [Online]. Available at: http://www.unece.org/fileadmin/DAM/trans/main/tem/temdocs/ Proposal_4_council_regulation_regime_border.pdf [accessed: July 22, 2013].

COM(2009)383 final. Commission of the European Communities. 2009. Report from the Commission to the European Parliament and the on the implementation and functioning of the local border traffic regime introduced by Regulation (EC) No. 1931/2006 of the European Parliament and of the Council laying down rules on local border traffic at the external land borders of the Member States. Brussels. [Online]. Available at: http://ec.europa.eu/anti_ fraud/documents/euro-protection/directive_conterfeiting_en.pdf [accessed: July 22, 2013].

COM(2011)47 final. European Commission. Communication from the European Parliament and the Council. Second report on the implementation and functioning of the local border traffic regime set up by Regulation No. 1931/2006. Brussels. [Online]. Available at: http://eur-lex.europa.eu/LexUriServ/LexUriServ.do? uri=COM:2011:0047:FIN:EN:PDF [accessed: July 22, 2013].

COM(2011)461. Commission of the European Communities. 2011. Proposal for a Regulation of the European Parliament and of the Council amending Regulation(EC) No. 1931/2006 as regards the inclusion of the Kaliningrad area and certain Polish administrative districts in the eligible border area. COM(2011)461. Brussels. [Online]. Available at: http://ec.europa.eu/home-aff airs/news/intro/docs/110729/1_EN_ACT_part1_v6.pdf [accessed: May 30, 2013].

Commission of the European Communities. 2013. Proposal for a Regulation of the European Parliament and of the Council establishing a Registered Traveller Programme. 2013/0059 (COD). Brussels. [Online]. Available at: http:// ec.europa.eu/dgs/home-affairs/doc_centre/borders/docs/1_en_act_part1_v14. pdf [accessed: May 30, 2013].

Consolidated Version of the Treaty of European Union. European Union. Consolidated Versions of the Treaty on European Union and of the Treaty Establishing the European Community. Official Journal of the European Union of 29 June 2006. [Online]. Available at: http://eur-lex.europa.eu/ LexUriServ/LexUriServ.do?uri=OJ:C:1997:340:SOM:EN:HTML [accessed: August 26, 2013].

Council of the European Union. 1999. The Schengen Acquis as referred to in Article 1(2) of Council Decision 1999/435/EC of 20 May 1999, Official Journal of the European Communities. [Online]. Available at: http://eur-lex. europa.eu/LexUriServ/LexUriServ.do?uri=OJ:L:2000:239:0001:0473:EN:P DF [accessed: July 22, 2013].

Council of the European Union. 2006. Council Regulation (EC) No. 1932/2006 of 21 December 2006 amending Regulation (EC) No. 539/2001 listing the third countries whose nationals must be in possession of visas when crossing the external borders and those whose nationals are exempt from that requirement. [Online]. Available at: http://eur-lex.europa.eu/LexUriServ/LexUriServ.do?uri =OJ:L:2006:405:0023:0034:EN:PDF [accessed: July 22, 2013].

European Parliament and the Council. 2011. Regulation (EU) No. 1342/2011 of the European Parliament and of the Council of 13 December 2011 amending Regulation (EC) No. 1931/2006 as regards the inclusion of the Kaliningrad oblast and certain Polish administrative districts in the eligible border area, Official Journal of the European Union. [Online]. Available at: http://eur-lex.

europa.eu/LexUriServ/LexUriServ.do?uri=OJ:L:2011:347:0041:0043:EN:P
DF [accessed: July 22, 2013].

European Parliament and the Council. 2006. Regulation (EC) No. 562/2006 of
the European Parliament and of the Council of 15 March 2006establishing
a Community Code on the rules governing the movement of persons across
borders (Schengen Borders Code). [Online]. Available at: http://eur-lex.
europa.eu/LexUriServ/LexUriServ.do?uri=OJ:L:2006:105:0001:0032:EN:P
DF [accessed: July 22, 2013].

European Union. 2006. Consolidated Versions of the Treaty on European Union
and of the Treaty establishing the European Community. Brussels. [Online].
Available at: http://eur-lex.europa.eu/LexUriServ/LexUriServ.do?uri=OJ:C:2
006:321E:0001:0331:EN:PDF [accessed: July 22, 2013].

European Union. 2010. Consolidated versions of the treaty on European Union
and the Treaty on the Functioning of the European Union (2010/C 83/01),
Official Journal of the European Union C 83/1, Official Journal of the European
Union C 81/1. [Online]. Available at: http://eur-lex.europa.eu/LexUriServ/
LexUriServ.do?uri=OJ:C:2010:083:FULL:EN:PDF [accessed: July 22, 2013].

Regulation (EC)1931/2006. European Parliament and the Council. 2006.
Regulation (EC) No. 1931/2006 of the European Parliament and of the
Council of 20 December 2006 laying down rules on local border traffic at the
external land borders of the Member States and amending the provisions of
the Schengen Convention. Regulation(EC) No. 1931/2006. Brussels. [Online].
Available at: http://eur-lex.europa.eu/LexUriServ/LexUriServ.do?uri=OJ:L:20
06:405:0001:0022:EN:PDF [accessed: May 30, 2013].

Further References

Altvater, E. and Mahnkopf, B. 2007. *Konkurrenz für das Empire*. Münster:
Westfälisches Dampfboot.

Apap, J., Carrera, S., and Kirişci, K. 2004. Turkey in the European Area of Freedom,
Security and Justice. CEPS-EU-Turkey Working Papers No. 3. [Online].
Available at: http://www.ceps.be/book/turkey-eu-area-freedom-security-and-
justice [accessed: August 26, 2013].

Balscok, I., Dancs, L., and Koncz, G. 2005. Bridge or Iron Curtain? Local
Hungarian and Ukrainian Perceptions of a New European Union Border.
Journal of Borderland Studies, 20(2), 47–66.

Balibar, E. 2009. Europe as Borderland. *Environment and Planning D: Society
and Space*, 27, 190–215.

Bekus-Goncharova, N. 2008. Living in Visa Territory. [Online]. Available
at: http:www.eurozine.com [accessed: November 30, 2009].

Belina, B. and Miggelbrink, J. 2013. Risk as a Technology of Power. FRONTEX
as an Example of the De-politicization of EU Migration Regimes. *The Spatial
Dimension of Risk: How Geography Shapes the Emergence of Riskscapes*,
edited by D. Müller-Mahn. London/New York: Routledge, 124–36.

Berg, E. and Ehin, P. 2006. What Kind of Border Regime is in the Making? Towards a Differentiated and Uneven Border Strategy. *Cooperation and Conflict*, 41(53), 53–71. [Online]. Available at: http://cac.sagepub.com/ content/41/1/53 [accessed: July 22, 2013].

Białasiewicz, L., Dahlman, C., Apuzzo, G.M., Ciut, F., Jones, A., Rumford, C., Wodak, R., Anderson, J., and Ingram, A. 2009. Interventions in the New Political Geographies of the European "Neighborhood." *Political Geography*, 28, 79–89.

Bigo, D. 2005. Frontier Controls in the European Union: Who is in Control? *Controlling Frontiers: Free Movement Into and Within Europe*, edited by D. Bigo and E. Guild. Hants (GB), Burlington (USA), 49–99.

Browning, Ch. and Joenniemi, P. 2008. Geostrategies of the European Neighbourhood Policy. *European Journal of International Relations*, 14(519), 519–51. [Online]. Available at: http://ejt.sagepub.com/content/14/3/519 [accessed: July 22, 2013].

Brunet-Jailly, E. 2006. Security and Border Security Policies: Perimeter or Smart Border? A Comparison of the European Union and Canadian–American Border Security Regimes. *Journal of Borderland Studies*, 21(1), 3–20.

Bruns, B. 2010. *Die soziale Organisation von Schmuggel am Rande der Europäischen Union* [Social Order of Smuggling at the Edge of the European Union]. Wiesbaden: VS Verlag.

Bruns, B., Miggelbrink, J., and Müller, K. 2011. Smuggling and Small-scale Trade as Part of Informal Economic Practices: Empirical Findings from the Eastern External EU Border. *International Journal of Sociology and Social Policy*, 31(11/12), 664–80.

Bruns, B., Miggelbrink, J., Müller, K., Wust, A., and Zichner, H. 2013. Making a Living on the Edges of a Security Border: Everyday Tactics and Strategies at the Eastern Border of the European Union. *Theorizing Borders through Analyses of Power Relationships*, edited by P. Gilles, H. Koff, C. Maganda, and Ch. Schulz. Brussels, P.I.E. Land, 89–110.

Buckel, S. and Wissel, J. 2009. Entgrenzung der europäischen Migrationskontrolle. *Demokratie in der Marktgesellschaft*, edited by H. Brunkhorst. Baden-Baden: Nomos, 385–403.

Carrera, S., Hernandez, N., and Parkin, J. 2013. Local and Regional Authorities and the EU's External Borders: A Multi-level Governance Assessment of Schengen Governance and "Smart Borders." [Online]. Available at: http://cor. europa.eu/en/documentation/studies/Documents/LRAs_and_EU_external_ borders/LRAs_and_EU_external_borders.pdf [accessed: July 22, 2013].

Cox, K. 2002. *Political Geography*. Oxford: Oxford University Press.

De Boer, K. 2010. Frontex: Der falsche Adressat für ein wichtiges Anliegen. *Kriminologisches Journal*, 42(3).

Donnan, H. and Wilson, T.M. 2003. Territoriality, Anthropology, and the Interstitial: Subversion and Support in European Borderlands. *Focaal—European Journal of Anthropology*, 41, 9–20.

Dubowski, T. 2012. Local Border Traffic—European Union and Member States' Perspective (based on Polish experience). *European Journal of European Law*, 14, 367–91.

Emerson, M. 2002. The Wider Europe as the European Union's Friendly Monroe Doctrine. CEPS Policy Brief N. 27. [Online]. Available at: http://www.ceps.be [accessed: July 10, 2013].

Foucault, M. 2007. *Security, Territory, Population: Lectures at the Collège de France, 1977–78*. Basingstoke: Palgrave Macmillan.

Haase, A., Wust, A., Knappe, E., and Grimm, F.-D. 2004. Wandel in ostmittel-europäischen Grenzregionen. Auswirkungen der zunehmenden Durchlässigkeit der polnischen Ostgrenze auf Grenzregionen und Grenzbeziehungen. Leipzig: Leibniz-Institut für Länderkunde.

Hapke, H. 2001. Petty Trader, Gender, and Development in a South Indian Fishery. *Economic Geography*, 77(3), 225–49.

Harvey, D. 2007. The Kantian Roots of Foucault's Dilemmas: Space, Knowledge and Power. *Foucault and Geography*, edited by J.W. Crampton and S. Elden. Aldershot: Ashgate, 41–7.

Heeg, S. and Ossenbrügge, J. 2002. State Formation and Territoriality in the European Union. *Geopolitics*, 7(3), 75–88.

Hohnen, P. 2003. *A Market out of Place? Remarking Economic, Social, and Symbolic Boundaries in Post-Communist Lithuania*. Oxford: Oxford University Press.

Holtom, P. 2004. Shuttle Trade and New Border Regimes. [Online]. Available at: http://www.iiss.org/rrpfreepdfs.php?scID=60 [accessed: August 5, 2013].

Huysmans, J. 2000. The European Union and the Securitziation of Migration. *Journal of Common Market Studies*, 38(5), 751–77.

Jileva, E. 2002. Visa and Free Movement of Labour: The Uneven Imposition of the EU Acquis on the Accession States. *Journal of Ethnic and Migration Studies*, 4, 683–700.

Jones, A. and Murphy, J.T. 2011. Theorizing Practice in Economic Geography: Foundations, Challenges, and Possibilities. *Progress in Human Geography*, 35(3), 366–92.

Kaiser, R. and Nikiforova, E. 2006. Borderland Spaces of Identification and Dis/location: Multiscalar Narratives and Enactments of Seto Identity and Place in the Estonian–Russian Borderlands. *Ethnic and Racial Studies*, 29(5), 928–58.

Kuus, M. 2007. *Geopolitics Reframed: Security and Identity in Europe's Eastern Enlargement*. New York: Palgrave Macmillan.

Lavenex, S. 2006. Shifting Up and Out: The Foreign Policy of European Immigration Control. *West European Politics*, 29(2), 329–350.

Lavenex, S. and Schimmelfennig, F. 2009. EU Rules beyond EU Borders: Theorizing External Governance in European Politics. *Journal of European Public Policy*, 16(6), 791–812.

Müller, K. and Miggelbrink, J. 2014. "The Glove Compartment Half-full of Letters"—Informality and Cross-border Trade at the Edge of the Schengen Area.

The Informal Post-socialist Economy. Embedded Practices and Livelihoods, edited by J. Morris and A. Polese. London and New York: Routledge, 152–64.

Müller, K. 2013. Vor den Toren der Europäischen Union. Handlungsorientierungen ökonomischer Akteure an der östlichen EU-Außengrenze. [Outside the Gates of the European Union. Action Orientation of Economic Actors at the Eastern External Border of the EU]. Wiesbaden: Springer VS.

Newman, D. 2006. The Lines that Continue to Separate Us. *Progress in Human Geography*, 30(2), 143–61.

Nikiforova, E. 2005. Narrating "Nationals" at the Margins: Seto and Cossack Identity in Russian–Estonian Borderlands. *Culture and Power at the Edges of the State: National Support and Subversion in European Border Regions*, edited by Th. Wilson and H. Donnan. Münster: Lit, 191–228.

O'Connell, K. 2008. EU-visa Policy: Squaring the Circle of Neighbourhood and JHA Objectives. *Europe's Near Abroad: Promises and Prospects for the EU's Neighbourhood Policy*, edited by D. Mahncke and S. Gstöhl. Brussels: P.I.E. Peter Lang, 115–34.

Paasi, A. 1998. Boundaries as Social Processes: Territoriality in the World of Flows. *Geopolitics*, 3(1), 69–88.

Painter, J. 2010. Rethinking Territory. *Antipode*, 42(5), 1090–118.

Papagianni, G. 2006. *Institutional and Policy Dynamics of EU Migration Law*. Leiden: Martinus Nijhoff Publishers.

Peers, S. 2011. *EU Justice and Home Affairs Law. 3rd Ed.* Oxford: Oxford University Press.

Polese, A. 2012. Who has the Right to Forbid and Who to Trade? Making Sense of Illegality on the Polish–Ukrainian Border. *Subverting Borders. Doing Research on Smuggling and Small-scale Trade*, edited by B. Bruns, J. Miggelbrink (Hrsg.), and S. Wiesbaden, 21–38.

Rijpma, J.J. and Cremona, M. 2007. The Extra-territorialisation of EU Migration Policies and the Rule of Law. EUI Working Paper LAW, No. 2007/01. Available at: http://papers.ssrn.com/sol3/papers.cfm?abstract_id=964190 [accessed: February 15, 2010].

Round, J., Williams, C.C., and Rodgers, P. 2008. Everyday Tactics and Spaces of Power: the Role of Informal Economies in Post-soviet Ukraine. *Social and Cultural Geography*, 9(2), 172–85.

Sack, R. 1983. Human Territoriality: A Theory. *Annals of the Association of American Geographers*, 73, 55–74.

Smith, A. and Stenning, A. 2006. Beyond Household Economies: Articulations and Spaces of Economic Practice in Postsocialism. *Progress in Human Geography*, 30(2), 190–213.

Taylor, P.J. 1982. A Materialist Framework for Political Geography. *Transaction of the Institute of British Geographers*, 7(1), 15–34.

Thuen, T. 1999. The Significance of Borders in the East European Transition. *International Journal of Urban and Regional Research*, 23(4), 738–50.

Van der Velde, M. and Van Naerssen, T. 2001. People, Borders, Trajectories: an Approach to Cross-border Mobility and Immobility In and To the European Union. *Area*, 43(2), 218–24.

Van Houtum, H. and Pijpers, R. 2007. The European Union as Gated Community: the Two-faced Border and Immigration Regime of the EU. *Antipode*, 37, 291–309.

Van Houtum, H. and Van Naerssen, T. 2002. Bordering, Ordering and Othering. *Tijdschrift voor Economische en Sociale Geografie*, 93(2), 2002, 125–36.

Van Schendel, W. and Abraham, I. 2005. *Illicit Flows and Criminal Things: States, Borders, and the Other Side of Globalization*. Bloomington: Indiana University Press.

Vaughan-Williams, N. 2008. Borderwork Beyond Inside/Outside? FRONTEX, the Citizen-detective and the War on Terror. *Space and Polity*, 12(1), 63–79.

Wagner, M. 2010. Die moralische Ökonomie des Schmuggels. *Alltag im Grenzland: Schmuggel als ökonomische Strategie im Osten Europas*, edited by M. Wagner and W. Łukowski. Wiesbaden: VS Verlag, 73–89.

Walters, W. 2002. Mapping Schengenland: Denaturalizing the Border. *Environment and Planning D: Society and Space*, 20(5), 561–80.

Wesseling, M. and Boniface, J. 2011. New Trends in European Consular Services: Visa Policy in the EU Neighbourhood. *Consular Affairs and Diplomacy*, edited by J. Melissen and A.M. Fernández. Leiden: Martnius Nijhoff Publishers.

Wust, A. and Zichner, H. 2010. "Here is the Wall!"—Is it? Transborder Practices of Small-scale Economic Actors at the Ukrainian–Romanian Border. *Revue d'études comparatives Est–Ouest*, 41(4), 171–93.

Xheneti, M., Smallbone, D., and Welter, F. 2012. EU Enlargement Effects on Cross-border Informal Entrepreneurial Activities. *European and Regional Studies*, 20(3), 318–28.

Ziltener, P. 2000. Die Veränderung von Staatlichkeit in Europa—regulations- und staatstheoretische Überlegungen. *Die Konfiguration Europas*, edited by H.-J. Bieling and J. Steinhilber. Münster: Westfälisches Dampfboot, 73–101.

SECTION III
Border Work by Non-Traditional Actors Away from the Border

Chapter 8

Symbolic Bordering and the Securitization of Identity Markers in Nigeria's Ethno-Religiously Segregated City of Jos

Yakubu Joseph and Rainer Rothfuss

Introduction

Borders have remained an integral part of human political and social life in stark defiance of the predictions at the dawn of the twentieth century that globalization would bring about the emergence of a borderless and deterritorialized world. Borders have not only retained their relevance in our world today, but are adapting to its evolving spatiotemporal conditions. Conventionally, borders delineated the spatial extent of state sovereignty and served mainly as static spatial device for "cordoning" a political territory against potential external threats. However, interdisciplinary border research has witnessed a paradigm shift from such state-centric national security focus to a decentralized society's (human) security concern. Here the state and the population are all active players and shapers of borders. The concept of borders as fixed demarcating lines that are space and time oriented, is now being replaced with one in which borders are defined in terms of bordering, an emphasis on the symbolic and social practices of spatial differentiation aimed at controlling movement of people both into and within a securitized space (Houtum and Naerssen 2002).

This shift to a more fluid and mobility focused border has increased all over the world in the wake of the September 11 terrorist attacks in the United States. Bordering is now rather employed as a mechanism for isolating and preventing a security risk personified by the Other, seen as a mobile threat that needs to be kept at bay within and without a securitized space. A growing interest in this phenomenon is already taking place (Jones 2012, Popescu 2012).

In this study, we examine bordering in the context of conflict within society. Specifically, the study explores how bordering, as a socio-spatial process, has been employed as a neighborhood security strategy in the ethno-religiously segregated city of Jos in Nigeria. The implications of the symbolic and constructive nature of spatial differentiation of "we" and the "Other" including the production and reproduction of an "enemy picture" through a bottom-up agency are significant. For the segregated residents of Jos, Debrix and Barder's (2009) assertion that there is nothing to fear but fear is instructive in understanding the basis and impact of this bordering (othering) strategy. In this chapter, we attempt to show how the fear of repeated violence and consistent failure of the state to deliver on its promise to guarantee the security of life and property of its citizens has led the Muslim and Christian communities in Jos to engage in mutual profiling and bordering practices based on socio-spatial differentiation of identities. Two key points are salient here. First, ordinary citizens do not leave their fate in the hands of the state when faced with the threat of inter-group violence and/or terrorism, but try to create barriers to keep the "bad guys" out. The second aspect of the argument of this chapter is that borders that separate the conflicting communities in Jos are not only physical but also discursive. Through border narratives and symbolization the two communities have been able to demarcate themselves into compartments of "us" and "them" as a way of minimizing the risk of attacks. Consequently, in the context of conflict, bordering acts as an enhancer of stereotypes and damaging xenophobia, which sustains mutual distrust and hinders the prospect of peace.

The Background of Ethno-religious Conflict in Jos

Jos is the capital of Plateau State, located in central Nigeria at the intersection of the country's largely Christian south and the predominantly Muslim north. As its name suggests, Plateau State sits on a plateau—with undulating hills and spectacular landscape—giving it a favorable climate with relatively low temperatures throughout the year compared to the rest of the country due to its high altitude. The state used to be known as a "Home of Peace and Tourism" due to its attraction to local and foreign tourists and its peacefulness. It has also been widely described as a microcosm of Nigeria because it is a melting pot of ethnic and religious diversities, though Christians are the overwhelming majority (Egwu 2004, Ambe-Uva 2010, Higazi 2011). A large number of traditional religious worshippers abound all over the state. Many historical factors that contributed to the cosmopolitan nature of Jos have been identified in numerous studies (Egwu 2004, Falola and Heaton 2008, Abdulkadir, 2011, Abbas 2012). These include the location of Jos as a trading transit route for several centuries, its role as a safe haven for those fleeing from the Usman dan Fodio led Fulani jihad from 1804 to 1808, and as a site for the influx of mining workers in the early parts of the twentieth century. Plateau State was an important mining site for the British colonial government.

By and large, the residents of Jos co-existed peacefully until the last part of the twentieth century (Ambe-Uva 2010: 42). Contestations of the ownership of Jos between two groups—the Berom, Anarguta, and the Afizere on one hand and the Hausa/Fulani on the other—was reinforced by the decision of the military junta of General Ibrahim Babangida to create and delineate Jos North Local Government in 1991, which was perceived by the former as a deliberate attempt to dispossess them of their land and give the latter, who are heavily concentrated in the area, a political advantage (Ostien 2009). Since then isolated clashes have occurred between the two groups around the question of who will serve as the Chairman of the Local Government. However, on September 7, 2001, there was a conflagration of violence in the city triggered by contestations over the appointment of a poverty eradication coordinator (Ostien 2009, Krause 2011). This confrontation between those referred to as indigenous ethnic groups (the Berom, Anarguta, and Afizere) and those referred to as settlers (Hausa/Fulani) snowballed into a religious conflict because the former are predominantly Christians and the latter are predominantly Muslims (Krause 2011, Higazi 2011). As a consequence, the conflict polarized the population and pitted the groups against each other along the indigene/settler and Christian/Muslim divides. Sadly, since 2001 the state lost its reputation as a Home of Peace and Tourism. Jos became the scene of pernicious cycles of violence that claimed thousands of lives (Krause 2011). Despite numerous peacebuilding initiatives, today the city is a shadow of its past image as a model of integration and remains highly polarized and segregated ethno-religiously. The word "border" has become part of the lexicon of daily life in Jos, not as a reference to Plateau State's or Nigerian boundary, but to the arbitrary and imagery lines of spatial differentiation between the conflicting parties.

Theoretical and Conceptual Framework

At the end of the Cold War, and with the fall of the Berlin Wall, globalization inspired among its many enthusiasts the anticipation that the world was moving toward becoming a borderless and deterritorialized place (Newman 2008: 133). This utopian borderless and deterritorialized world was meant to represent the triumph of peace over conflict, inclusion over exclusion, and unity over division. The scenario presupposed that the co-dependence and interconnectedness of human societies will overcome the need for fences and demarcation in the world. Contrary to this, as Newman (2006: 143) asserted: "We live in a world of lines and compartments. We may not necessarily see the lines, but they order our daily life practices, strengthening our belonging to, and identity with, places and groups, while—at one and the same time—perpetuating and reperpetuating notions of difference and othering." While some borders might have disappeared, or are disappearing, many new ones are being established (Newman 2008). In general, instead of being phased out, borders remain pervasive in today's world. Jones (2012) notes that even countries considered as consolidated democracies have built physical barriers and the creation of securitized spaces in recent years. In a nutshell, ours has become "A Barricaded World" (Blij 2009: 28).

As reality starkly contradicts the notion of a borderless and deterritorialized world, the study of borders and boundaries has continued to flourish among human geographers and scholars from across many disciplines (Newman 2006, Newman 2008 and Moraczewska 2010). Since 9/11 there has been an unprecedented increase of interest in borders (Popescu 2012). This also means that there is a growing repository of theories and perspectives on the subject. Much has been written about borders and their spatial and social implications for international relations (Moraczewska 2010), state, and society (Blij 2009); international peace and security (Jones 2012); borderlands (Zartman 2010); and politics of identity (Newman 2008) among many other aspects. The traditional preoccupation of political geography with borders and boundaries had focused largely on the demarcation of the spatial extent of the state, that is, the delineation of the geographical area within which the state exercises sovereignty (Newman 2008: 124). Although this traditional interest still permeates the study of borders and boundaries, contemporary studies are focusing more on the dynamics of borders and the changing patterns in which people and groups relate "within a variety of social and spatial compartments" (Newman 2008: 128). Newman further observes a new wave of interest in the study of borders, which focuses on the discursive and symbolic dimensions. This chapter falls within this discursive and symbolic realm of political geographers' interest in borders. Looking at borders beyond being mere physical barriers and demarcations of securitized spaces, we examine how border narratives and the symbolization of borders mediate in the construction of identity and sustain the ethno-religious conflict divide in Jos. In their article, "Bordering, Ordering and Othering," van Houtum and Naerssen (2002: 125) underscore the growing academic interest in the study of "practices of ordering and the discursive differentiation between us and them, seen through the lens of spatial bordering," since the 9/11 terrorist attacks in the United States.

In order to explain the practices of discursive construction of identity with its corollary "us" versus "them" dichotomy that have become the order of the day in Jos, we would first attempt to establish the philosophical and theoretical underpinnings of the nexus between bordering and security and how it relates to this context. We begin by providing a theoretical foundation for understanding the involvement of non-state actors in security maintenance. This allows the reader to appreciate the conditions that have prompted ordinary residents in Jos to engage in profiling the "other" and social-spatially demarcating their neighborhoods by means of border narratives and discursive construction of identities.

In his famous classical work, *Leviathan*, Thomas Hobbes (1588–1679) canvassed for a social contract in which a potential *Bellum omnium contra omnes*, i.e. "war of all against all," may be prevented through the exercise of force by a central authority. The Hobbesian model dreads the "state of nature," characterized by civil disorder, and demands citizens to yield their propensities for violence to a sovereign, the state, who should monopolize the use of force in order to prevent the society from drifting towards anarchy. The Hobbesian model found some resonance with the conceptualization of the "state of exception" by the German philosopher, political theorist, jurist, and law professor, Carl Schmitt (1888–1985). Schmitt equated sovereignty with the degree to which the state mobilizes a condition of exception. Thus, for him, "the legal and constitutional order of the modern state must have the capacity to cast itself out through an act or decision of the executive, but also potentially through the actions of policing or soldiering agents, during a period necessary to defeat a deadly enemy" (Debrix and Barder 2009: 399). As Debrix and Barber further noted, the takeaway here from the Hobbesian "mechanism of ensuring control over fear/danger" and the Schmittian "condition of sovereign exception" have elevated fear into the political realm and defined security over a territorialized space in terms of how the source of fear or danger is controlled.

Borders have become means by which the state territorializes fear and securitizes spaces. In the wake of 9/11, we have seen states increasingly employ the mechanisms of centralization of fear (Debrix and Barder 2009) through stronger securitization of borders (Moraczewska 2010, Popescu 2012). In tandem with this securitization of borders, the very paradigm of security has changed. Popescu (2012) identified the salient features of this change—a shift from national security primarily concerned with the defense of territorial sovereignty to an understanding of security as more connected with daily life. Consequently, the binary distinction between national security and personal security is altered—the former is no longer viewed as merely external concern, to be handled by the military, and the latter not just a domestic concern to be taken care of by the police. Thus, instead of a preoccupation with how to guard against external threats to state territory, securing everyday life has become part of the norm (Dillon 2007 cited in Popescu 2009). As a result, the border has become operationalized as a mechanism for the management of risks to everyday life through its presumed selective permeability, having the ability to prevent incursion by threats (or unwanted strangers) into securitized spaces (see Bosworth 2008). This development has been the subject of numerous studies on security sector reforms around the world in numerous fields, including political geography, peace and security studies, legal studies, and international relations. The privatization of security and border control has increased significantly since 9/11. Private contractors have exploited the growing opportunities offered by the increased securitization of borders including the application of surveillance devices such as biometric technologies (see Schreier and Caparini 2005, Popescu 2009).

The implication of the paradigm shift in the conception of security highlighted above is that risks are defined as "existential threats to the identity of a social group" and that of the security of everyday life of the individual, and a lot of "people feel they experience these threats without the mediation of the state" (Popescu 2012: 92). While classical approaches to national security, which are often obsessed with territorial sovereignty, operate on the basis of the centralization of fear, the new realities of society's security have made fear dispersed and decentralized among populations. Consequently,

> [...] borders have emerged as society's security guarantors and have uncritically become part of everyday life.

> However, borders are more than risk management sites. They are security constitutive as well. Border-making discourses play active parts in the production of societal and human security risks, for it is at the crossing of a border that someone or something can become a security risk. Othering continues to take

centre stage in these discourses, with the caveat that the Other has shifted from the neighboring nation to mobile phenomena. Put differently, the blurring of the inside/outside territorial distinction characteristic of modern states has not led to the disappearance of this border-based power practice. Borders continue to provide the basis for inside/outside differentiation with regard to group membership. What has changed is the type of territorial logic involved in Othering, which has moved now beyond fixity to include flexibility and multiplicity, that is, network membership. (Popescu 2012: 93)

The significance of the change highlighted by Popescu is that it underscores at once the twin processes of bordering and othering. As a potential threat is defined in terms of a mobile risk that seeks to invade a securitized space, discursive bordering lends itself to the strategy of spatial differentiation of insiders (us) and outsiders (them). These dynamics render themselves more intelligible when bordering processes are interrogated rather than the fixed lines of separation. Van Houtum and Naerssen (2002: 126) have, more than a decade ago, made a compelling observation in this regard:

Bordering processes do not begin or stop at demarcation lines in space. Borders do not represent a fixed point in space or time, rather they symbolise a social practice of spatial differentiation. Semantically, the word 'borders' unjustly assumes that spaces are fixed in space and time, and should rather be understood in terms of bordering, as an ongoing strategic effort to make difference in space among the movements of people, money and products.

Therefore, insiders and outsiders are not only differentiated at the port of entry, that is, at the point of crossing demarcating lines, but also often through the spatial construction of identities within a securitized space. Within a given space an individual or a group may be discursively cataloged as intruders and be treated as potential security risks for a society. Spatial perceptions play a mediating role in this process of othering. Glassner and Fahrer (2004) elaborate the concept of spatial perception. They note that it is shaped by cultural conditioning, values, attitudes, motivations, and goals. People develop cognitive maps of places which are imagery, idealized, and stereotyped. On the basis of such mental maps, they construct in-group and out-group identities. Social cognitive theories of stereotyping offer insights into this human enigma and the social construction of reality in relation to social and power relations in society (Augoustinos and Walker 1998). Sociologists have attempted to expand the concept of social stereotype beyond the experience of the individual to the level of social group. Susan Condor analyzed the relationship between social stereotypes and social identities. According to her, social stereotypes can be viewed as: Images formed of and formed by human beings; cultural constructs; shared beliefs; communicative acts; and implicated in social structure (Condor 1990). Due to these characteristics, stereotypes are able to serve the purpose of bordering and othering. The stereotypes built about a people are represented by symbols and narratives for ease of differentiating in-group and out-group members, making it possible to entrench fear as well as inclusion and exclusion over a space.

In his lecture series, *Society Must Be Defended*, Michel Foucault advanced the concept of biopolitics. He drew a contrast between the biopolitics of fear in the modern state and the classical concept of the politics of sovereign exceptionality. While the politics of sovereign exceptionality is concerned with the fear and exercise of power over a political territory, biopolitics of fear, on the other hand, has shifted the focus from the political territory to the population. By means of governmentalized techniques and procedures concerned with the regulation of a population, fear is redistributed by means of a biopolitical arrangement (Debrix and Barder 2009). Central to the relationship between biopolitics and security is the fraught notion of biopower, a power that focuses on the population and its main preoccupation is to ensure proper governance of life in society. The applications of power across society have thus become increasingly a matter of dealing with differences in a population and not mere territories. To explain the relations of biopower to bordering we turn to Louise Amoore's work. According to Amoore (2006: 338), "Subject to biopower, the crossing of a physical border is only one crossing in a limitless series of journeys that traverse and inscribe the boundaries of safe/dangerous, civil/uncivil, legitimate traveler/illegal migrant." By virtue of biopower, the body thus has, in effect, become the carrier of border. Amoore contends that states have the wherewithal to establish and maintain border management security to extend the regulating of mobility into multiple aspects of daily life through biometric profiling of multiple encoded borders—social, legal, gendered, and racialized borders. This has become the growing trend since 9/11 and as a result of the global war on terror.

This chapter examines how ordinary citizens have exercised biopower in dealing with insecurity, engendered by ethno-religious segregation. The apparent failure of the state, either by omission, commission, or incapacity, to fulfill what Schmitt would call its sovereign prerogative by preventing pernicious cycles of ethno-religious violence, has given rise to social groups and individuals assuming the responsibility of securing the borders of their segregated, confined neighborhoods. With neither the legal right nor the wherewithal to enforce biometric checks on the movement of people into their fear-defined enclaves, the communities use symbols and discourses to construct their own identities as insiders and outsiders that are seen as constituting a security risk to everyday life. Hence, constructed attributes are assigned to the Other. Risk is, therefore, narrowly defined as a probability that threats will be kept at bay by disallowing anyone whose appearance, behavior, or mannerism invokes the attributized enemy image. As discussed in the following section, the conflict in Jos has shown that the process of border formation, which used to be top-down and at the territorial level of the state, is also exercisable by bottom-up agency at the grassroots level.

The Making of a Segregated City

The city of Jos witnessed eruptions of large scale inter-communal violence in 2001, 2002, 2008, and 2010, which started as a clash between "indigenes" and "settlers" and metamorphosed into a protracted Muslim–Christian conflict (Higazi 2011: 2). Also, between 2010 and 2013 there were a number of terrorist bomb attacks targeting public spaces and churches by the Islamist militant group Boko Haram as well as guerilla style attacks on farming villages around the environs of Jos. The attacks by Boko Haram included the use of suicide bombers and planting of Improvised Explosive Devices in Christian neighborhoods. Generally, violence was geographically dispersed across the city and beyond. Some neighborhoods were not as severely affected as others. It also became clear that the possibility of escaping the carnage was a function of one's location. Many victims of the early confrontations were vulnerable because they lived in neighborhoods where their group was a minority or got caught up in interface areas or common places. This observation does not ignore the reality that many people were killed or lost their homes and belongings even in areas they considered less vulnerable. A lot of people travelling through the city had also been killed because Jos is a main transit hub that links the country's South and the North.

The first wave of segregation began in the aftermath of the 2001 crisis. There are suggestions that many residents moved out of neighborhoods they felt were unsafe to those they deemed safe (see Best and Rakodi 2011). With every episode of violence the city became increasingly bifurcated along a Christian–Muslim divide. Even though the conflict has ethnic and political dimensions, it appears that religion was the most potent identity for mobilizing people and creating the strongest sentiment. For example, the Pew Forum found that the majority of Nigerians—91 percent of Muslims and 76 percent of Christians—consider religion as the most important identity and more significant than their identity as Africans, Nigerians, or members of an ethnic group (Few Forum on Religion and Public Life 2006). As residents felt that the security forces were not able to prevent the violence and keep them safe, communal security initiatives began to emerge. Vigilante groups were sporadically established in different neighborhoods. These vigilante groups, comprising predominantly of youths and supervised and coordinated by elders,[1] mounted makeshift roadblocks along the perimeters of and all access roads to their neighborhoods. They searched cars and screened people coming into the area.

License plates are one way these vigilantes patrol neighborhoods for people they think do not belong. Because the plate carries the abbreviations of the registering State and also a Local Government Council, it has become fashionable for people to have plate numbers of their home states and Local Government Councils. Sometimes people prefer registration numbers of certain states or Local

---

1   Interview with a Muslim community leader, who was working with his Christian counterparts to monitor the vigilante group in Federal Low Cost area in Jos.

Government Councils even though they may not reside in those places. As car registration numbers become stereotyped as suggesting a person's ethnic origin, Jos residents monitor the movement of cars within their neighborhoods by their plate numbers. Having a plate number associated with the Other's state of origin when there is tension in the city may affect one's movement across the segregated city or even put someone at risk of being lynched by a mob.

The activities of these vigilante groups remained very controversial. They may be mobilized as militia groups either for self-defense or for reprisal attacks (Best and Kemedi 2005). Best and Kemedi's study provides great insights into the role of vigilante groups and militias in the Jos conflict. Although the police are making efforts to regulate the activities of vigilante groups[2] their spontaneous, informal, and scattered nature makes it difficult to enforce code of conduct on their operations.

The "borders" inadvertently established by the activities of the vigilante groups and the relocation of people to places they consider as safe havens for their religious communities determine the contours of the socio-spatial segregation that has occurred in the city. As fear continuously becomes part of the security architecture of the communities, the securing of the separating lines, usually road networks, between the conflicting parties was seen as insufficient. To compound this situation, the terror attacks against churches by the Islamist group Boko Haram[3] added another thick layer of fear on the people's minds. Such attacks against churches were usually followed by reprisals against Muslims by irate Christian youths. Boko Haram has been engaged in an insurgency against the Nigerian State, killing Christians, bombing churches, killing fellow Muslims, and attacking security posts and schools in a bid to create an Islamic state in northern Nigeria (Amnesty International 2013). Increasingly, fear and mutual suspicion characterize the relationship between the Christian and Muslim communities. The two communities resorted to new processes of bordering to prevent possible infiltration of their space by the Other. Christians erected fences and gates, and mounted barriers against potential suicide bombers who might want to crash improvised explosive device loaded cars into churches to harm worshippers. Parking lots were designated in places far from churches. Muslims also erected barricades around their mosques especially during the Jummat prayers. Public buildings and private businesses took measures to secure their building through the use of security barriers and metal detectors.

Security discourses among residents began to shift the emphasis on borders to asking people to be vigilant and report any suspicious movements to the authorities

---

2   Inauguration of a vigilante group in Jos, *News Agency of Nigeria*, March 8, 2013. Available at: http://ww.nanngronline.com/picture/inauguration-of-a-vigilante-group-in-jos [accessed: July 7, 2013].

3   The name Boko Haram means "Western education is an abomination" in Hausa language. The group's real name is *Jamā'a Ahl al-sunnah li-da'wa wa al-jihād* (People Committed to the Propagation of the Prophet's Teachings and Jihad).

Figure 8.1   A church and school compound in Jos surrounded by barriers to prevent bombers from getting close

Note: Photo by Yakubu Joseph, September 2012.

or community leaders in their neighborhoods. This gave rise to profiling and othering across the segregated neighborhoods. For example, Nigeria's highest ranking security officer, the Chief of Defense Staff, Air Marshall Oluseyin Petinrin, on a visit to the troubled city of Jos, called on the residents to be vigilant and security-conscious.[4] Suspicion of any strange looking person may cause panic and irrational reaction from the community. For example, a mentally ill person was spotted in the Rantya area of the city and security agents were called in to apprehend him because residents thought he was a disguised terrorist. When his items were searched the security agents discovered that he was just a mentally ill

4   Petinrin asks Jos residents to be vigilant, *Daily Trust*, December 14, 2011. Available at: http://www.dailytrust.com.ng/index.php/news-news/12002-petinrin-asks-jos-residents-to-be-vigilant [accessed: July 7, 2013].

person picking up garbage. The fear of the Other remained a part of the daily life in Jos. A recent story carried by the *This Day* newspaper reported that a group calling itself Vigilant Democrats, based in Jos, raised an alarm and alerted residents of the restive city about the presence of 30 suspected foreign assassins brought in by the enemies of peace in the state to destabilize the city and to wreak havoc on citizens.[5]

Symbolic Bordering and the Securitization of Identity Markers in Jos

In this section we analyze the different categories of symbolic and constructed identity markers employed in the spatial differentiation of insiders and outsiders across the city. As the mutual suspicions of the Other mediates the way people perceive security, bordering was adopted as a strategy of gatekeeping to prevent the invasion of "own" space by the Other. Consequently, the construction of identity through spatial differentiation became entrenched in the security discourses. The "life and death" determining question, *"Namu ne ko nasu ne?"*[6] (literally meaning "are you for us or for them?") used by both sides against strangers during the violence was normalized as part of everyday life. The symbolization and discursive construction of identities in space became well packed with stereotypes and derogatory expressions that continue to nurture demonizing perceptions and hostility towards the Other. Such spatial perceptions are accompanied by a mental mapping, that is, ordering and othering of the constructed markers of identity. Residents of Jos cast themselves into Christian and Muslim spatial identities based on constructed and taken for granted signifiers. In her work on the Jos conflict, Jana Krause (2011: 10) concluded that "Ten years of violent confrontations and the extreme brutality of 2010's massacre around Jos left many residents traumatized. Religious identities have become strongly polarized and one-sided narratives internalized." This is a challenge for co-existence, as Kuna (2005: 11–12) points out, the relationship between identity and religion is profound:

> People's perceptions and definitions of themselves as adherents of specific religious experiences, in a sense 'in-groups,' against outsiders that do not share the same religious experience or 'out-groups,' and the way in which institutional systems provide each with readily available repertoires that make life that much meaningful are critical elements in the construction of identities. [...] To a very significant extent therefore, religion shapes identity not just in specifying physically recognizable 'signifiers,' but also by imposing notions of membership

---

5   "Group Raises the Alarm Over Suspected Foreign Assassins in Jos," *This Day*, April 28, 2013. Available at: http://www.thisdaylive.com/articles/group-raises-the-alarm-over-suspected-foreign-assassins-in-jos/146148/ [accessed: July 7, 2013].

6   This is an expression in Hausa language. Hausa is the Lingua Franca spoken widely across northern Nigeria including Plateau State.

within a community of believers sharing common beliefs and values, whose lives are that much more organized with, than without religion.

The negative impact of this act of bordering is quite complex. Even for in-group members this is a dicey business because the Nigerian society has loose compartments of identities and is socio-culturally defined by its vast diversities. Therefore, the probability of mistaking people's identities in some instance is very high. Nigeria has more than 250 ethnic groups. Many of these cultural and linguistic groups exhibit a high degree of mutual intelligibility (cf. Ballard 1971).

Jos residents have adopted physical attributes as identity markers to establish spatial differentiation between those who belong to a place and those who do not. Physical attributes refer to how a person looks in terms how he or she is built, facial shape, skin color, hair type and style, height, and so on. Like racial profiling, residents of Jos try to differentiate their identities by assigning physical attributes to the Other. Hence, physical attributes are stereotyped and used as identity markers. A stranger bearing the appearance attributed to that typical of the Other is treated in an unwelcoming and hostile manner or even likely to be subjected to scrutiny and harassment within a neighborhood. This bordering practice of determining the religious identity of a person based on his or her physical attributes in order to spatially differentiate insiders and outsiders undermines the demographic realities of Nigeria. It is foolhardy to assume that someone's physical attributes would clearly reveal his or her ethnic identity, and one's ethnic identity would suggest a person's religious affiliation (cf. Best and Rakodi 2011: 23). Unfortunately, this fallacy has been imbibed by many residents, as a result of the politicization of fear, in order to close their spaces to the Other. There were reported instances of people harassed for mistaken physical attributes by members of their own religious group. Two persons interviewed during the study recounted how they were almost killed during clashes by members of their own religious group because of their physical attributes in neighborhoods where they were not known.

Language has also become a key marker of religious identity. For example, a middle age woman who met one of the authors (Joseph) in Jos at a car wash a few years ago asked him to serve as a translator between her and a cobbler (a mobile shoe repairer). After Joseph had successfully facilitated the transaction by translating between English and Hausa, the woman felt at home with him and began to share her personal experience directly. Although she belongs to the Igbo ethnic group from southeastern Nigeria, she was born in Jos and lived there all her life. That revelation instantly brought a flicker of shock and bewilderment into Joseph's mind. How on earth would someone born and brought up in Jos claim not to understand the Lingua Franca of the region? Impossible, Joseph thought in his mind. He politely asked her, "Ma, does it mean you understand Hausa language?" The lady replied, "Yes, I used to speak Hausa, but after my family lost everything in the crisis in Jos I decided never to speak that language again." The woman's refusal to speak Hausa after an experience of personal tragedy as a result of the ethno-religious crisis in Jos shows her perception about the language identity of the Other. In many neighborhoods in Jos, the language one speaks is regarded as a clue of the person's religious identity. There is a backlash against Hausa, which is the Lingua Franca in the North including Plateau, from the conflict as demonstrated by the lady in the story. This phenomenon had been documented by Higazi (2011: 4): "There has been a reaction against this [use of Hausa as a Lingua Franca] in some areas, with a cultural resurgence that is encouraging the replacement of Hausa place names and ethnonyms with indigenous ones, and to a lesser extent the use of indigenous languages rather than Hausa." How a person speaks a language is also labeled. Speaking a language with the accent of a native speaker or a non-native speaker could make a person be identified with a particular ethnic group and by implication be specified a religious identity.

Clothing serves as a third cultural marker of religious identity to distinguish between insiders and outsiders across neighborhoods in Jos. It has been a trend in Nigeria for parents to dress their children in traditional ceremonial attire during festivals or to pose for a photograph on their birthdays. The photos of children dressed in such traditional outfits are a common part of interior decor of many homes. During recent field work, one of the authors (Joseph) visited a family in Jos. The family's living room was decorated with photo frames of important family milestones—such as the couple's wedding, naming ceremonies, and birthdays. Joseph remarked to his hosts that the photo of their ten-year-old son dressed in a legendary traditional dress was beautiful. The response from the parents was not a thank you, but a shocking revelation of the toll ethno-religious conflict has taken on culture, and on how stereotypes serve as the barbed wire of bordering. The parents said their little boy is not happy with those photographs and had refused to wear those dresses ever since he became aware of the othering, the polarization between indigenes and settlers, Christians and Muslims, engendered by conflict. For the young boy, that traditional dress is associated with the Other. The parents had to get rid of three pairs of that expensive garment because their boy had developed a dislike for wearing them.

Part of the cultural resurgence referred to by Higazi (2011) include the adoption of certain dress codes and rejection of others. Some ethnic groups have distinct traditional dressing, albeit many of them share common traditional dress. Since one's mode of dress can be associated with a particular religious affiliation, in the different neighborhoods of Jos residents maintain dressing codes acceptable to their groups. Moving with the "wrong" mode of dress in a neighborhood may attract a hostile reaction from residents. As residents remain "vigilant and security-conscious," their symbolic bordering "telescopes" can easily detect the strange appearance of the Other in their securitized space. As a result of this othering, both Christians and Muslims disguise with the mode of dress of the Other when they are entering the Other's territorialized neighborhoods or interface areas to avoid the risk of being harassed or even attacked.

The potency of the dress code as religious identity marker was further revealed when a top-ranking, female public servant described her encounter with a powerful politician at a government office in Jos.[7] A meeting was scheduled and she was to be in attendance. While she was waiting for a quorum to be formed, the politician came in. As soon as he saw how she was dressed he remarked to the people he knew there, "Why are you guys parleying with the Taliban?" To contain the embarrassment, one of the persons there said to him, "Honorable, she is one of us." The lady has been part of many peace initiatives in the state and still maintains interactions with her Christian friends. Her family still lives in one of the very few mixed neighborhoods in Jos. During her daughter's wedding, she even made her Christian friend the "Mother of the Day," a big honor in Nigerian wedding celebrations. She felt extremely bad for being cast in a negative light simply because of the way she dressed. Many of these experiences are reported across the different neighborhoods in Jos.

Her story illustrates the fourth cultural marker of religious identity: an individual's network of friends. Bordering in the segregated neighborhoods of Jos has assumed a vicarious dimension. By associating with the Other, a person risks having his or her loyalty to one's religious affiliation questioned. Since the Other is considered an enemy to be kept at bay, any in-group members that associate with the enemy are seen as exposing their community to risk or betraying the hate the Other deserves. This othering practice, in which those who seek to cross the borders of hate and fear, and associate with fellow human beings across the conflict divide are viewed with suspicion and disapproval, has been very difficult for many people who can no longer visit their childhood friends, schoolmates, former neighbors, and colleagues at home. Friendship between men and women from different faiths is frowned upon. Prior to the crisis intermarriage was common between Christians and Muslims, but these days this can even trigger restiveness in the city. For example, there had been reported incidences of violence sparked by protest over a dating relationship between young men and young women from the two religions (Human Rights Watch 2005: 14).

The final key cultural markers of religious identity in Jos are names. Many Christians and Muslims in northern Nigeria bear common names because of common ethnicity and the use of Hausa as a Lingua Franca throughout the region. Due to the common Abrahamic roots of Christianity and Islam, and the Arabic etymological foundations of many Hausa words, Hausa names of the prophets and ancient religious personalities were commonly used by Christians and Muslims. However, as a result of the conflict and the strong need each side feels for a clear differentiation of identities, Christians have developed a preference for the English versions of biblical names and words that have special meanings. Some of the names that have become trendy among Christians nowadays are Grace, Peace, Faith, Miracle, Divine, Favor, Promise, and Prevail. Muslims, on the other hand, prefer the Arabic and Hausa variants of common Arabic names including Mohammed, Aldulsalam, Fatima, Aisha, Mubarak, and Suleiman. Many daily

---

7   This was a personal account given by the lady during an interview in her office in 2011 in Jos.

life decisions that are taken concerning individuals are mediated by how the individual's name is perceived as a Christian or Muslim. The newcomer seeking to rent an apartment, the young man or young lady applying for a job, and the vendor trying to sell a product are all subjected to name filtering to determine whether they deserve a favorable or negative response.

Conclusion: Bordering and the Challenge of Restoring Peace in a Divided City

This chapter examined spatial segregation and the relationship between bordering and security during a period of conflict in the city of Jos. The inability of the state to prevent the recurrence of violence in the city for the past 12 years and the traumatic impact of the crisis on residents prompted communal initiatives of securitizing spaces across the ethno-religiously polarized city. As the fear of possible invasion by the Other spread throughout the city, both conflicting communities adjusted their conceptions of security from securing the borders of their territorialized social and spatial compartments from perceived external existential threats to spatially differentiating the movement of insiders and outsiders across and within socio-spatially constructed boundaries. This phenomenon has underscored the notion that borders are not simply the fixed demarcating lines between spaces (van Houtum and Naerssen 2002).

Constructed symbolic markers of identity are the "borders" that determine the spatial differentiation of insiders and outsiders along ethno-religious lines in Jos. License plates, language, names, clothing, physical attributes, and even networks of friends are used as cultural markers of religious identity to reinforce stereotypes and sustain an "enemy picture" of the Other. Although bordering and othering practices have helped to somewhat keep the belligerents asunder and to offer the communities the semblance of security without eliminating the underlying fear, at the same time they pose a serious challenge to the parallel efforts by many stakeholders to bridge the divide between the communities in order to bring about healing and reconciliation, and to build sustainable and durable peace in Plateau State. Therefore, in the context of this conflict, bordering is an ambivalent process. It is a two-edged sword that ostensibly reduces security risk to a social group or a territorialized space and at the same time produces socio-spatial outcomes such as stereotypes, hate, and adversarial posturing. Understanding these spatial and territorial dimensions of the Jos conflict will contribute to the development of an integrative security framework (cf. Newman 2008: 134) in which the subjective concerns of both sides would be merged into an inter-subjective goal of creating security for everyone.

The implication is that the nexus between securitization of spaces and peacebuilding must be adequately considered in designing a new architecture of peace in Jos. As this study has shown, bordering can be exercised by both state and non-state actors and can create social and material borders far away from an international border line. In a situation of conflict, where the state has failed to demonstrate the capacity to tackle security threats effectively, belligerents and ordinary community members may resort to enacting socio-spatial barriers to their perceived enemies. These socio-culturally constructed and context specific socio-spatial barriers are arbitrary and fluid, rendering themselves to abuse and, even though they serve an immediate security need, they may hinder the prospect for reconciliation and peace. It is therefore important for scholars to pay attention to this phenomenon of symbolic and discursive bordering with its attendant othering consequences.

References

Abbass, I.M. 2012. No Retreat No Surrender Conflict for Survival Between Fulani Pastoralists and Farmers in Northern Nigeria. *European Scientific Journal*, 8(1), 331–46.

Abdulkadir, M.S. 2011. Islam in the Non-Muslim Areas of Northern Nigeria, c. 1600–1960. *Ilorin Journal of Religious Studies*, 1(1), 1–20.

Ambe-Uva, T.N. 2010. Identity Politics and the Jos Crisis: Evidence, Lessons and Challenges of Good Governance. *African Journal of History and Culture*, 2(3), 42–52.

Amnesty International 2013. Annual Report 2013: The State of the World's Human Rights. London: Amnesty International. Available at: http://files.amnesty.org/air13/AmnestyInternational_AnnualReport2013_complete_en.pdf [accessed: July 7, 2013].

Amoore, L. 2006. Biometric Borders: Governing Mobilities in the War on Terror. *Political Geography*, 25, 336–51.

Augoustinos, M. and Walker, I. 1998. The Construction of Stereotypes within Social Psychology: From Social Cognition to Ideology. *Theory Psychology*, 8(5), 629–52.

Ballard, J.A. 1971. Historical Inferences from the Linguistic Geography of the Nigerian Middle Belt. *Africa: Journal of International African Institute*, 41(4), 294–305.

Best, S.G. and Kemedi, D.V. 2005. Armed Groups and Conflict in Rivers and Plateau States, Nigeria. *Armed and Aimless: Armed Groups*, edited by N. Florquin and E.G. Berman. Geneva: Small Arm Surveys, 13–45.

Best, S.G. and Rakodi, C. 2011. Violent Conflict and its Aftermath in Jos and Kano, Nigeria: What is the Role of Religion? International Development Research, University of Birmingham, Religions and Development Research Working Paper No. 69.

Blij, H.D. 2009. *The Power of Place: Geography, Destiny, and Globalization's Rough Landscape*. New York: Oxford University Press.

Bosworth, M. 2008. Border Control and the Limits of the Sovereign State. *Social and Legal Studies*, 17(2), 199–215.

Condor, S. 1990. Social Stereotypes and Social Identity. *Social Identity Theory: Constructive and Critical Advances*, edited by D. Abrams and M. Hogg. London: Harvester Wheatsheaf, 230–48.

Debrix, F. and Barder, D. 2009. Nothing to Fear but Fear: Governmentality and the Biopolitical Production of Terror. *International Political Sociology*, 3(4), 398–413.

Dilloon, M. 2007. Governing Through Contingency: The Security of Biopolitical Governance. *Political Geography*, 26, 41–7.

Egwu, S.G. 2004. Ethnicity and Citizenship in Urban Nigeria: The Jos Case, 1960–2000, [Doctoral Dissertation]. Available at: http://dspace.unijos.edu.ng/handle/10485/850 [accessed: July 8, 2013].

Falola, T. and Heaton, M. 2008. *A History of Nigeria*. New York: Cambridge University Press.

Few Forum on Religion and Public Life. 2006. Spirit and Power: A 10-Country Survey of Pentecostals. Available at: http://www.pewforum.org/2007/03/21/nigerias-presidential- election-the-christian-muslim-divide/.

Gibbs, G.R. 2007. *Analyzing Qualitative Data*. London: SAGE Publications.

Glassner, M. and Faher, C. 2004. *Political Geography*, 3rd Ed. Hoboken, NJ: John Wiley and Sons.

Higazi, A. 2011. The Jos Crisis: A Recurrent National Tragedy. [Online: Friedrich Ebert Stiftung], Discussion Paper No. 2, January.

Houtum, H.V. and Naerssen, T.V. 2002. Bordering Othering and Othering. *Tijdschrift voor Econnomische en Sociale Geografie*, 93(2), 125–36.

Human Rights Watch 2005. *Revenge in the Name of Religion: The Cycle of Violence in Plateau and Kano States*, 17(8(A)). Available at: http://dspace.cigilibrary.org/jspui/bitstream/123456789/20153/1/Revenge%20in%20the%20Name%20of%20Religion%20The%20Cycle%20of%20Violence%20in%20Plateau%20and%20 Kano%20States.pdf?1.

Jones, R. 2012. *Border Walls: Security and the War on Terror in the United States, India and Israel*. London: Zed Books.

Krause, J. 2011. *A Deadly Cycle: Ethno-religious Conflict in Jos, Plateau State, Nigeria*. Geneva: Geneva Declaration.

Kuna, M.J. 2005. Religion, Identity, and National Integration in Nigeria. National Institute for Policy and Strategic Studies Round Table on National integration in Nigeria, Kuru, Jos, Nigeria, July 15 2005. Available at: http://academia.edu/255519/Religion_Identity_and_National_Integration_in_Nigeria [accessed: July 7, 2013].

Moraczewska, A. 2010. The Changing Interpretation of Border Functions in International Relations. *Revista Română de Geografie Politică*, 12(2), 329–40.

Newman, D. 2006. The Lines that Continue to Separate Us: Borders in our "Borderless" World. *Progress in Human Geography*, 30(2), 143–61.

Newman, D. 2008. Boundaries. *A Companion to Political Geography*, edited by J. Agnew, K. Mitchell, and G. Toal. Malden, MA: Blackwell Publishing, 123–37.

Ostien, P. Jonah Jang and the Jassawa: Ethno-Religious Conflict in Jos, Nigeria. Muslim Christian Relations in Africa. [Online]. Available at: http://papers.ssrn.com/sol3/papers.cfm?abstract_id=1456372 [accessed: July 2, 2013].

Popescu, G. 2012. *Bordering and Ordering the Twenty-first Century: Understanding Borders*. Plymouth: Rowman and Littlefield Publishers.

Schreier, F. and Caparini, M. 2005. Privatising Security: Law, Practice and Governance of Private Military and Security Companies. Geneva Centre for the Democratic Control of Armed Forces, Occasional Paper No. 6. Available at: http://dspace.cigilibrary.org/jspui/bitstream/123456789/27442/1/Privatising%20Security%20%20Law,%20Practice%20and%20Governance%20of%20Private%20Military%20and%20Security%20Companies.pdf?1 [accessed: July 5, 2013].

Zartman, W. 2010. *Understanding Life in the Borderlands: Boundaries in Depth and in Motion*. Athens, Georgia: University of Georgia Press.

Chapter 9

Border Wars: Narratives and Images of the US–Mexico Border on TV

Reece Jones

The Cable Television Wars

In 2013, the United States cable television landscape is a place of war. Animal Planet has "Whale Wars," which documents environmentalists' efforts to disrupt Japanese whaling operations. There is a "Star Wars" remake on Spike TV. Spike TV also has another show called "The Deadliest Warrior" in which different teams of soldiers face off in competitions. Over on the Travel Channel, "Food Wars" serves up battles between different restaurants to make the best version of a particular dish. Episode titles include "Philly Cheese Steak War" and "Chicago Pizza War." The Discovery Channel has "Weed Wars," which documents the lives of people who run medical marijuana dispensaries. HGTV, the Home and Garden Network, has "Design Wars," in which designers "battle it out" to design rooms in a house. A & E, formerly the Arts & Entertainment Channel, has four war shows. "Parking Wars" follows parking enforcement officers who give tickets to illegally parked cars. "Storage Wars" glorifies people who buy the contents of abandoned storage units at auctions. "Storage Wars Texas" just does it bigger, because everything is bigger in Texas. "Shipping Wars" follows independent truckers who ship odd-sized items. The Food Network probably wins the war of having the most unlikely war show with "Cupcake Wars." The show's website includes a graphic of a large pink cupcake with a tank gun protruding out of it. With all of these other vacuous uses of the term "war," you cannot blame the National Geographic Channel for calling a show about US Border Patrol agents using helicopters, unmanned drones, and machine guns on the US–Mexican border "Border Wars." Nevertheless, in a nod to the particularly American banalization of war, when the show is broadcast on most National Geographic stations around the world, "war" is dropped from the title and it becomes simply "The Border."[1]

"Border Wars" was an immediate success and its first episode on 10 January 2010 was the highest rated premier ever for the National Geographic Channel. The show is in its fifth season and is still in production. "Border Wars" utilizes what appears to be a documentary style and follows the experiences of Border Patrol agents and Customs officers over several shifts on the job. The narratives and images in the

---

1   One exception is in Australia where it is called "Mexican Border Wars."

show are often the first time many viewers see the Border Patrol and what occurs in the borderlands. "Border Wars" takes the unknown space of the border and transforms it into a series of images and stories that create a coherent narrative for the viewer. The show is a powerful propaganda tool that portrays the Border Patrol as brave, patriotic, and compassionate as they simultaneously fight the war on drugs, battle terrorism, and save the lives of immigrants stranded in the desert. The show does not, however, put these fragments in the context of why people cross the border, why they choose such a difficult route through the desert, or where the confiscated drugs are going.

This chapter analyzes the representation of the US–Mexican border in National Geographic's "Border Wars." Representations and narratives play a critical role in shaping perceptions of chaotic and distant events. We cannot be everywhere at once and we cannot know what is occurring over a vast area. Anderson (1991) argues newspapers standardized accounts across a wide readership, which allowed people to share in the knowledge of events in places far distant from their daily life. Before newspapers, limited first-hand experience and unreliable word-of-mouth produced fragmented and varied accounts of events. Erving Goffman (1979: 27) contends that visual mediums can be even more effective than text in shaping our understanding of events because images "transform otherwise opaque goings-on into easily readable form." This transformation of the opaque into perceivable knowledge is very powerful and consequently, as Castells (2010: xxxii) writes, "power struggles have always been decided by the battle over people's minds, this is to say, by the management of processes of information and communication that shape the human mind."

Critical geopolitics analyzes the social construction of the political world by investigating the narratives and actors that create representations of geopolitical space (Dodds 2001, Ó Tuathail 1996). Rather than accepting a fixed reality in the world, the focus is instead on how perceptions of reality are created for particular purposes. These invented worlds can entail both representations of territories and representations of people, which constitute the effort to categorize and define the subjectivity of an individual or a group. Once established and inscribed into the consciousness of a population, these geopolitical discourses act as disciplinary regimes of truth by shaping how events are understood and interpreted by the population (Foucault 1971, 2002). Consequently, defining the boundaries of the categories we use to understand the world defines what is and is not (Jones 2009).

After situating the recent increases in manpower and budget of the Border Patrol within the history of the US–Mexico border, this chapter examines the first season of "Border Wars" through the lens of critical geopolitics and identifies five reoccurring themes that shape the image of the border and the Border Patrol for the viewer (Fairclough 1995, Müller 2010).[2] These themes are the presumption of guilt,

---

2   This paper is a critical geopolitical analysis of the representations of the US–Mexico border in the show border wars. The first season of the show was analyzed by transcribing each episode and applying the lens of critical discourse analysis to the narrative

the potential for violence, the language of war, the lack of governance in Mexico, and the simultaneous dehumanization of the immigrants and valorization of the Border Patrol agents. What emerges through this border work by the producers of "Border Wars" is a sharp disconnection between, on the one hand, the framing of each segment, the language of the narrator, and the perspectives of the Border Patrol agents, and on the other hand the footage of what actually happens at the border in each episode. Despite the militaristic lead-ins, the dramatic music, and the tension of the storytelling that emphasizes violence, terrorism, and war, most of the episodes present a more prosaic border landscape peopled by poor migrant workers looking for a better life. The question remains, however, whether the viewer remembers the dramatic and frightening set-ups or the banal denouement when another group of immigrants is rounded up, hand-cuffed, and put in the back of a Border Patrol truck.

"What it is Really Like"

Over the past 30 years, the US–Mexico border entered the US political debate as a symbol and touchstone for understanding a range of changes occurring in society (Andreas 2009, Heyman 1998, Nevins 2010). The border was described as a bridge for trade in the form of Maquiladoras and later NAFTA and as a dangerous space that needed to be secured to protect American jobs from immigrants and American children from the scourge of drugs. The attention to the border occurred during a period when there were profound changes in how the border was monitored and patrolled, which resulted in substantial increases in funding for the Border Patrol (Ackleson 2005, Coleman 2003, 2005, Dunn 2009, Heyman and Ackleson 2009, Jones 2012, Lytle Hernandez 2010). From 1980 to 1995, the Border Patrol budget increased sevenfold. From 2000 until 2010 its budget tripled again increasing from $1.06 billion to $3.58 billion (Haddal 2010).

A large portion of the budget increases went to hiring additional agents and fencing the border. In 1992, at the US–Mexico border there were 3,555 agents and by 2010 there were over 20,100 (Haddal 2010).[3] Additionally, in 2006, the US Congress passed the Secure Fence Act that authorized fencing on 1,100 km of the border of the 3,169 km border, and 1,070 km were completed by 2010 (Haddal et al. 2010). These mutually constitutive processes of hiring agents while constructing material barriers to movement create a new landscape as the border becomes securitized and militarized.

National Geographic's "Border Wars" is significant because it brings the hidden and often opaque borderlands and the activities of the Border Patrol into the homes

---

representations of the show (Fairclough 1995, Müller 2010). CDA attempts to identify the ways that power operates through the narrative construction of reality.

3   There are an additional 1000 agents US–Canada border, however over 98 percent of apprehensions occur at the Mexico border.

of millions of Americans, and viewers around the world, every week. The show is shot in a documentary style, but it often depicts the most spectacular aspects of the work of Border Patrol agents and Customs officers. "Border Wars" uses images of Predator drones, Black Hawk helicopters, hidden seismic sensors, and night vision equipment to build excitement and tension in the show. The publicity material for the show and its producer, Nicholas Stein, emphasize that the show strives to show the real experiences of the border.

In a 2010 television interview, Stein explains:

> We were there to pull back the curtain and let people see exactly what it's like day to day, car by car, mission by mission, shift by shift, what it's really like to try to secure the US–Mexico border. And in many ways Nogales became a microcosm, if you will, of some the issues and problems that are up and down all the way from San Diego all the way to Brownsville, Texas. So it's a real look at the work and the dedication of the men and women there. We didn't talk policy, we didn't talk about, you know, what people should do in terms of policy and legislation and laws. We were there with the law enforcers and we saw how difficult their job really is. (Cavanaugh and Heilbrunn 2010)

The show is described as a documentary about the border; however, Stein explicitly states that he sees his job as telling the story from the perspective of the Border Patrol agents. He continues:

> There was an original show that National Geographic did called 'Border Wars' that was a one-hour show that was done by their Explorer Unit and it was sort of a—more of an overview of all the things that go on there.[4] But, really, this [Stein's version] is a look from the point of view of the federal law enforcement folks. There is, I think, many opportunities for many filmmakers of every stripe and news organizations to do a more comprehensive look at all of the issues there. There's so many points of view. But we decided that a lot of people really didn't understand what these men and women are being asked to do on our behalf and with our tax dollars. And we thought that it was important to get on the ground and really see that happening. The truth is there's no border like the US–Mexican border in the world because there's no border that has perhaps the world's richest country hard up against one of the poorest and one that's now going through the spasms of this narco war. (Cavanaugh and Heilbrunn 2010)

---

4   This previous documentary style show is referred to as Episode 1 of Season 1 of the show. The differences in style and content are clearly evident, as Stein suggests. This National Geographic Explorer documentary shows life on both sides of the border and it even follows an immigrant from Mexico all the way to his home in Kentucky. However, as is described later in this paper, it also uses similar production effects to create tension and drama in the storytelling.

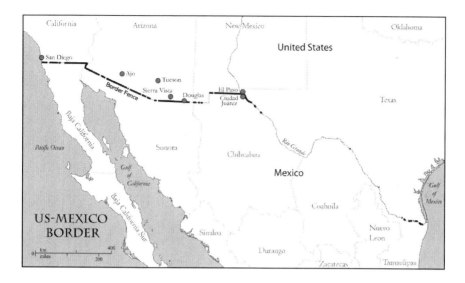

Figure 9.1   Map of US–Mexico border showing walled or fenced locations

The positive depiction of the Border Patrol is not surprising because the producers, and National Geographic, rely on the permission of the Department of Homeland Security and on the cooperation of the agents in order to film it. Stein explains the agreement with Customs and Border Protection,

> I think the CBP, Customs and Border Protection, *really trusts us to tell their story* in a serious way and to tell it in an accurate way. So after the negotiations were successful, we headed to Nogales, Arizona, and we got to know the officers and agents quite well. (Cavanaugh and Heilbrunn 2010, emphasis added)

Without that access, and the stock footage of the military hardware, the show would be impossible to make.

Fighting the Border Wars

Each episode of "Border Wars" begins with a fast-paced title sequence with images of guns, helicopters, Predator Drones, and agents racing through the desert accompanied by dramatic music and a voice-over about terrorism, drug cartels, security threats, and war. The first segment of each episode sets up several different scenarios involving Border Patrol agents on the ground, Rapid Response Team agents in a Black Hawk helicopter, and Customs officers checking cars and pedestrians at crossing points. The initial segment builds tension by emphasizing the potential threat through phrases that are repeated in virtually every episode about "trails known to be used by narco-traffickers," the fact that "smugglers are almost always armed," "will do anything to protect their cargo," and "ambushes are not uncommon." The remainder of the episode then follows the scenarios to their conclusion.

The title sequence and voiceover for the second episode of the first season is typical [description of the on-screen images and background music in brackets]:

> [Dramatic music] These officers and agents of the Department of Homeland Security work around the clock protecting America's borders [images of agents, ATVs, Black Hawk helicopters, and Predator drones]. They are at ground zero of the war against narco-traffickers, illegal immigration, and terrorism [dark, grainy film of a line of agents shooting automatic weapons after one yells 'fire']. In the next 12 hours, officers and agents in Nogales, Arizona will risk it all to pursue and arrest hundreds of illegal immigrants [image of an agent diving and tackling a man sitting on a hillside, the agent says: 'I was able to dive and get a hold of him just in time'], confront cartel foot soldiers in the dead of night [nighttime images of an agent with a gun pointed in front of him yelling 'Border Patrol! Parate (stop)! Parate!'] and stop would-be imposters from entering the country [image of a woman and an ID card]. These are the Border Wars. [Final image of several agents hiking past on a trail and one says: 'Welcome to Nogales'].

The sequence is exciting and frightening. The images do make it seem like a war is happening: there are agents in military fatigues, automatic weapons, military helicopters, Predator Drones, and video of what appears to be a firing squad. It also creates the perception of imminent threat at the border where agents "risk it all." The border is described as "ground zero of a war" against "narco-traffickers" and "terrorism" which includes "cartel foot soldiers" and "imposters" trying to slip into the country. An agent says he tackled a man "just in time," which implies that something terrible could have happened.

If you watch the sequence a few times, however, questions start to arise. All the accoutrements of war are there, but only one side. The guns, helicopters, agents, and Predator Drones are all from the US Border Patrol. "Terrorists," "narco-traffickers," and "cartel foot soldiers" are mentioned but none are shown. The Mexican military is not shown. The two threats that are shown are the imposter with the fake ID and the man tackled on the ridge "just in time." The imposter, as the episode later describes, is a middle-aged Mexican woman with three kids. She lived in the United States for 12 years before returning to Mexico to visit her dying father. Now she is trying to return to her job in California. The man tackled on the hill was a late-middle-aged undocumented worker. He was travelling with a large group of immigrants that was located by a helicopter. They ran in different directions and most of his group was caught by agents on the ground. This particular man was later seen by the helicopter pilot who aimed a spotlight on him so the agents on the ground could find him. The man did not run or resist. Instead, he sat still on the hillside in the spotlight for at least 10 seconds (the amount shown in the footage) before the agent dove at him from above and the two men rolled 50 meters down the steep incline. The older, slightly overweight man was later shown complaining to the agents about pain in his chest and legs after the fall. This pattern of exaggerating threats along the border is evident in every episode of the show.

In addition to the dramatic title sequences, there are several recurring themes in each episode that create tension but also establish a framework for viewers to understand what happens at the border. These themes are the presumption of guilt, the potential for violence, the lack of governance in Mexico, the language of war, and the dehumanization of immigrants and valorization of the Border Patrol agents. These representations support a particular narrative about the border that emphasizes threat and danger while providing little context to what is actually occurring in each episode. Indeed, just as the threats in the title sequences did not live up to the hype, the on-screen images throughout the show often belie the overwrought storytelling of "Border Wars."

The Presumption of Guilt

The first recurring theme in the show is the tendency of the producers, and the agents themselves, to presume that most of the people they encounter are, at the minimum, in the United States illegally and possibly are hardened criminals with violent intentions. This presumption is evident in how the narrator describes the people who are interdicted. Typically, the first time the people are mentioned they are referred to as "suspected" smugglers or illegal immigrants. However, later in the episode the terminology of suspicion is dropped or the narrator uses general terms like "smugglers are known to." Although the statements are about smugglers generally, the implication is that the people being shown on screen are examples of it.

A segment in the third episode of season one illustrates the tendency of agents to presume the people they detain are in the country illegally and more than likely criminals. The segment begins by showing Agent Pittman driving along the border fence in Nogales:

> Narrator: Pittman has patrolled this area for nearly 15 years and watched the violence intensify.

> Agent Pittman: I think it's more and more criminals coming across. It's every kind of criminal you can imagine we catch and arrest them.

> Narrator: Agent assaults are on the rise, including near the border fence.

> Agent Pittman: In my opinion it seems to be organized, trained people. They have set up ambushes for us. They would shoot at our guys.

Narrator: The biggest risk here are the rocks, bricks, and even kitchen sinks thrown over the fence. Cartels hire local teens to target the agents.

Agent Pittman: I still get nervous parking next to the fence.

A call comes over the radio that sends him to a remote area 15 kilometers from the city, which Pittman says "is [a] busy area for smuggling out here. In an emergency situation, everyone goes until we get enough people there. We have our agents being shot at a lot. We have cartels in Mexico telling their people to defend their loads at all costs."

The segment occurs during the daytime and Agent Pittman decides to pursue the suspects on foot, who are said to be walking along a dirt road that follows a gas pipeline. Pittman spots two men walking along the road in front of him and at the same time two agents on ATVs speed in and arrest them. The two 50- to 60-year-old men of apparently Latin American ancestry are wearing nice, clean clothing and carry themselves in a dignified manner. Despite their appearance, Agent Pittman carefully searches and interrogates them. First, he inquires if they have weapons or drugs. They respond "no." Then he asks if they are terrorists. Again, "no." Agent Pittman explains his concerns:

> They look like unarmed immigrants but the rule is never assume. It's dangerous—anything that can be made into a weapon like toothbrushes, combs, and pens, we'll take. Lighters, perfumes, that's flammable. We don't know who we are dealing with. They may be just looking for an opportunity to do something to harm you. We don't know their history, their criminal records until they get processed. You'd be surprised; we can't relax on these individuals because a lot of them do have criminal records.

Agent Pittman's approach to the two men is sanctioned by US law, which states that the Border Patrol can stop anyone near the border with "articulable facts" that led to their suspicion.

Section 287 (a)(3) of the Immigration and Nationality Act, 8 U.S.C. § 1357, authorizes Border Patrol agents, without a warrant, "Within a reasonable distance from any external boundary of the United States, to board and search for aliens any vessel within the territorial waters of the United States and any railway car, aircraft, conveyance, or vehicle." The Border Patrol has the authority to enter private property (but not buildings) within 25 miles (40 km) of the border. The reasonable external boundary distance for warrantless search and seizure was set as within 100 miles (161 km) of a land border or the coastline. In 2011, the US House of Representatives considered a bill that would further provide the Border Patrol with a waiver of 36 federal laws in the 100-mile zone near the borderlines (but not the coasts) in order to patrol for illegal entrants.

In the 1975 Brignoni-Ponce decision, the US Supreme Court reaffirmed the right of the Border Patrol to stop cars and pedestrians without warrants in the

100-mile zone under certain conditions "only if they are aware of specific articulable facts, together with rational inferences therefrom, reasonably warranting suspicion that the vehicles contain aliens who may be illegally in the country." The ruling goes on to list factors that could be considered "articulable facts:"

> Officers may consider the characteristics of the area in which they encounter a vehicle. Its proximity to the border, the usual patterns of traffic on the particular road, and previous experience with alien traffic are all relevant ... They also may consider information about recent illegal border crossings in the area. The driver's behavior may be relevant, such as erratic driving or obvious attempts to evade officers can support a reasonable suspicion ... The vehicle may appear to be heavily loaded, it may have an extraordinary number of passengers, or the officers may observe persons trying to hide. ... Aspects of the vehicle itself may justify suspicion. For instance, officers say that certain station wagons, with large compartments for fold-down seats or spare tires, are frequently used for transporting concealed aliens ... The Government also points out that trained officers can recognize the characteristic appearance of persons who live in Mexico, relying on such factors as the mode of dress and haircut. ... In all situations the officer is entitled to assess the facts in light of his experience in detecting illegal entry and smuggling.

Although the ruling limits the ability of Border Patrol agents to stop anyone, the factors listed are broad enough that virtually any stop could be justified (Heyman 2009). As the Supreme Court case indicates these articulable facts can be simply being near the border or wearing Mexican-style clothing or having a Mexican-style haircut.

Every episode of "Border Wars" demonstrates the result of these laws and court judgments as the Border Patrol agents operate with the normal American legal standard of presumption of innocence reversed. Instead, as Agent Pittman puts it, "the rule is" to treat everyone they encounter as a potential threat until they can prove otherwise.

The Potential for Violence

The Border Patrol reports that several hundred agents are attacked in the borderlands every year.[5] It is a dangerous job, particularly given the money involved in drug smuggling and human trafficking. In the show, the suspects are always set up to be potentially violent. With dramatic music in the background, the narrator states in ominous tones that the suspects are on a route that is known for smuggling, smugglers are almost always armed, attacks against agents are on the rise, and they will do anything to protect their cargo.

---

5   In 2009 the Border Patrol reported 1,073 attacks on agents.

A segment that follows the Black Hawk based Rapid Response Team in the fourth episode of season one demonstrates how the producers intentionally create fear through unequivocal statements about the dangerous threat posed by smugglers. The narrator begins (emphasis added):

> Oscar Peru and his team touch down in some of Arizona's *most inhospitable terrain*. Their goal: help ground agents capture what could be a group of dangerous smugglers without being attacked. Hundreds of agents are assaulted on the job each year. ... *Only the most experienced smugglers dare to navigate these dense woods*, relying on the harsh terrain to diminish their chances of apprehension.

The voice-over states that not only are they certain to encounter smugglers, but moreover it is only the most experienced smugglers due to their location.

After following tracks through the grove of trees, the agents finally apprehend the group at the end of the episode. Once the group is spotted, the show always cuts to a commercial break, heightening the drama. On many occasions, the final sound before the commercial resembles an echoing gunshot, which implies the agents were under fire or had to use force themselves. After the commercial break, however, it is revealed to simply be a group of immigrants. The narrator: "Not drug smugglers but half a dozen exhausted immigrants." The images show several men, two women and two small children both less than 10 years old. The problem with the documentary style of the show is evident in this encounter. The producers know the result of the search in the forest will be a family with small children, but they still state in a breathless voice that it is "most inhospitable terrain" and only "experienced smugglers" would dare to enter the woods.

The stories about ambushes and armed drug smugglers protecting their loads at all costs are equally misleading. In every episode of the first season when the agents encounter drug smugglers the men toss the loads and run as soon as the agents approach. Rather than hardened criminals, or "cartel foot soldiers," as they are referred to in the show, they are poor, desperate people who run away as soon as there is a chance of being arrested. Nevertheless, even as the agents collect the packages of marijuana, the voice-over continues to build tension stating "The smugglers could be waiting for the right moment to attack. Some will go to any lengths to protect their loads" (Episode 1.2). It is always "some" or "most" but apparently never the ones in the show.

Finally, in addition to stating that every trail depicted in the show is a "known smuggling route," the show indicates that virtually every situation increases the risk for the agents. In episode three of season one we learn that being near the border is more dangerous: "After dark the rules change, drug traffickers can ambush agents without warning and many will be armed. The closer you get to the border, the more brazen they get." However, episode four of season one tells the viewer that the farther into the United States a group gets the more dangerous they are: "The closer a group gets to safety, the more they are willing to risk and the more dangerous it is for agents on their tail." The reality is that every episode in every location is represented as being extremely dangerous and risky whether in the end the agents locate drug mules, undocumented workers, families with kids, or even a cow that set off the seismic sensors.

The Lack of Governance in Mexico

Just as the show dehumanizes immigrants by depicting them as shadowy law breakers who are a potentially violent threat to agents, it also represents Mexico as an ungoverned place where the state lacks the ability to control its population or enforce its laws. The first episode of the season includes a segment in which the camera crew travels to the Mexican side of the border in order to document how immigrants reach the borderline.[6] The segment begins with dark images shot from a car covertly driving at night around the town of Altar in the Mexican state of Sonora. The voiceover says in ominous tones:

> This is a world few outsiders see run by a shadowy network that offers a range of services. Phony US papers and IDs, van rides to the border, guides, [and] pickups on the US side. If there is a police presence in Altar, our film crew never detected it. We did notice we were followed and observed at every turn, though it wasn't clear by whom. It's no secret what this town's main business is [camera zooms in on a shop selling backpacks]. We found shops selling backpacks, shoes, water bottles and caffeine pills on nearly every corner.

This first scene presents Altar as a seedy, dangerous place where the Mexican government is absent and order is enforced by the cartels that control the streets. The actual footage does not quite live up to the narrative that accompanies it. Indeed, the evidence of lawlessness is a shop selling backpacks, shoes, and water bottles. These first furtive images of Altar make it appear that the crew is secretly filming because it is too dangerous to do so in the open. However, the next few minutes of the episode include interviews with people in public during the daytime ranging from a vender selling backpacks to an immigrant planning to cross into the United States.

After these public interviews, the tone then shifts again. Mood music begins in the background and the daytime moving images are replaced by a series of snapshots that resemble those taken surreptitiously by a police team on a stakeout. They are in black and white and each new image is accompanied by the clicking sound of a camera. These production effects imply something illegal was filmed and could only be seen through these secretive means. The voice-over explains: "They are bundled into groups and packed 30 at a time into shuttle vans that will take them to the border. The vans line up in broad daylight. Everyone in town

---

6   As Nicholas Stein states in the above interview, this first episode was made by the National Geographic Explorer unit, not Stein and his production team.

knows where they are headed." The people in the vans may indeed cross the border, but driving someone to the border within the territory of Mexico is no more illegal than taking one of the four daily Greyhound buses from Tucson to Nogales in Arizona. Essentially the film crew found a town that caters to immigrant workers by providing supplies and a ride to the border, but the show presents it as if it is shocking and illegal. It is unclear however what the producers would expect the Mexican government to do in this situation.

The fifth episode of the first season is entitled "City Under Siege" and focuses on the differences between Nogales, Arizona and Nogales, Sonora. The episode begins "Nogales, Arizona: a border town under siege" [image of a vehicle with a machine gun mounted on the back]. A siege is the military blockade of a city with the intent of conquering it. It is simply not an accurate description of Nogales because the immigrants and smugglers moving through the area have no intention of conquering the city or even staying there. Instead they want to move past it as quickly as possible. However, "siege" creates a useful narrative that represents the US side as the victims of the border. The narrator continues:

> The twin cities share the same name, but little else. South of the fence the scene is one of stark poverty and severe overcrowding. With a population that is as many as 10 times as large, many are desperate to come north to the other side. The fence doesn't stop the people who try crossing over every day. It merely slows them down.

The episode describes the two sides of the border as "different worlds" three times, which positions the US as part of the modern, civilized, and orderly world and Mexico as not. In this version of the border, it is the critical line to prevent those uncivilized and potentially violent practices from entering the United States. The siege metaphor contributes to the desired image of barbarians at the gate attempting to overrun civilization.

In the show, spillover violence from Mexico is treated as an unambiguous fact as each episode emphasizes the disorder in Mexico and how it threatens the stability of the United States. A supervisor at the port of entry explains that he views the violence in Mexico as a threat on the US side: "Although much of the drug violence occurs in Mexico, it spills across the border into the US. There is a lot of danger over here. At my house, I have a double deadbolt and I have six dogs." The officer states matter-of-factly that there is spillover violence at the border but there is no evidence of it. In 2013, the US Congressional Research Service reported that:

> U.S. federal officials have denied that the increase in drug trafficking-related violence in Mexico has resulted in a spillover into the United States, but they acknowledge that the prospect is a serious concern. ... CRS is unable to develop fact-based conclusions about trends in drug trafficking-related violence spilling over from Mexico into the United States. (Finklea 2013: 1)

The cities of Juárez, Mexico and El Paso, Texas provide the strongest evidence against the claim of spillover violence. Juárez, Mexico has been devastated by the drug violence. In 2008 there were 1,600 murders, in 2009, 2,600 murders, and in 2010, over 3,000 murders, which makes Juárez one of the most dangerous cities in the world. El Paso is directly beside Juárez on the other side of the border and the Rio Grande. Despite its proximity to the violence, El Paso only had five murders in 2010, its lowest total in 47 years. In terms of overall crime rate, El Paso was the safest city with a population over 500,000 people in the United States in 2010 (El Paso 2011).

Despite the US government data on the lack of spillover violence and the visual evidence on the show that most people who cross the border are unarmed immigrants looking for work, "Border Wars" creates the perception that Mexico is an ungoverned territory with an uncivilized population that does not respect the rule of law. These ungoverned and uncivilized people represent a potentially violent threat to both border towns like Nogales and to the civilized way of life in the United States as a whole, which must be sealed off with a wall and patrolled by whatever means are necessary, including the latest military technologies developed for the battlefields of Iraq and Afghanistan.

Militaristic Language

On paper, the role of the Border Patrol changed dramatically after the 9/11 attacks in the United States. While previously immigration and smuggling were the main focus, after 9/11 terrorism prevention was elevated to the first priority of the agency (US Customs and Border Protection 2005). As security practices were reorganized, the Border Patrol was moved to the Department of Homeland Security and the Border Patrol's guidelines were rewritten to emphasize its role in preventing terrorism. The Border Patrol National Strategy (2005), while conceding that the vast majority of people are "economic migrants," argues that an "ever present threat exists from the potential for terrorists to employ the same smuggling and transportation networks, infrastructure, drop houses, and other support and then use these masses of illegal aliens as 'cover' for a successful cross-border penetration." In 2004, Customs and Border Patrol Commissioner Robert Bonner said after 9/11, "US Customs and Border Protection became the nation's first line of defense against terrorist threats" (US Customs and Border Protection 2004).

The show emphasizes the role the agents and officers play in preventing terrorism. When agents find a tunnel under the border fence in the fifth episode of season one, an agent remarks: "It could be for anything from drugs to terrorists and weapons of mass destruction. We don't leave tunnels be." A customs officer describes his duties at the border by saying "other than fighting terrorists," he looks for smuggled drugs and fake identifications. Episode 4 introduces a Customs officer who searches vehicles crossing the border as a member of the "Anti-Terrorism Contraband Enforcement Team."

The terminology of war also pervades the show. Beyond the war metaphor in the title and the siege metaphor for border towns, the show consistently uses militaristic language to describe the activities of the Border Patrol agents and the Customs officers. Customs and Border Protection is part of the Department of Homeland Security and is a law enforcement agency, not part of the military. The agents and officers are the equivalent of police officers, not soldiers. Their role is not to defend the United States from attack but rather to patrol the border for violations of immigration and customs laws. Despite the official focus on terrorism as priority number one for Customs and Border Protection, in practice, most agents and officers spend their time as they had before chasing immigrants in the desert and searching for drugs at ports of entry. Or sitting bored in their trucks (Marosi 2011). Nevertheless, in "Border Wars" the narrator and the agents and officers consistently describe their activities using the language of war.[7]

The first segments of most of the episodes in the first season include aerial footage of the deserts along the border, which the narrator describes as a "battlefield." The fourth episode of Season 1 shows Border Patrol agents beginning their shift; the narrator states "It's 8 am and a new shift begins at Border Patrol headquarters in Nogales. Agents gear up to face illegal immigrants, drug traffickers and terrorists [images of agents distributing machine guns]. Their battleground is 1,100 sq. miles of unforgiving desert."

While the Border Patrol agents are out in the field pursuing people and tracking them through the desert, the Customs officers at the port of entry are described in similar terms. In episode five of the first season:

> Narrator: At the port of entry, a new group of officers steps up to their mission. Take on the hundreds of drivers who cross the border every day ... Supervisor Mark Shanley also prepares his troops for battle.

> Officer Shanley: Yesterday was a kick-ass day. A lot of good seizures, a lot of good imposters. Number one, always remember officer safety is paramount. Be aware of your surroundings. The violence in Mexico is still there, it is still a threat. All right? Let's be safe out there.

The language simply does not match the duties of these officers. For the most part, they check the passports and ID cards of legitimate border crossers who are going shopping or to work on the other side of the border. It is a routine, mundane job that does not resemble a war or a battle in any way.

The language and many of the images of "Border Wars" create the perception that the Border Patrol agents and the Customs officers are part of the military and are fighting a war. Despite the fact that the duties of the agents and officers are still law enforcement, the practice of border security has been militarized over

---

7   Indeed, the Fox News channel and website puts the heading "America's Third War" on any stories that refer to the Mexican drug cartels or the border.

the past 20 years. This includes the deployment of the military along the border in form of National Guard troops and the acquisition of military technologies developed for wars abroad in Iraq and Afghanistan. These technologies include guns, surveillance technologies from night vision to high tech sensors, Blackhawk helicopters, and unmanned Predator Drones. As the wars abroad wind down, it appears that military technology industry is focused on converting these products to border security uses (Gregory 2011).

Indeed, "Border Wars" producer Nicholas Stein was one of two keynote speakers at a major security industry convention in 2011. The report about the speech in the organization's magazine noted the priceless publicity the show provides for the industry:

> He and his crew captured border patrol agents, ICE investigators and Coast Guard officers employing a wide range of security equipment—including helicopters, patrol aircraft, night vision equipment, mobile surveillance vehicles, Predator Drones, X-ray machines, all-terrain vehicles, body scanning equipment, portable fingerprinting devices and much more—as they portrayed the real-life challenges confronting U.S. Government personnel along the nation's southern and northern borders. ... The ratings for *Borders Wars* has been extraordinarily high, Stein theorized, 'because of the enormous hunger our viewers have for a real sense of what is going on down there.' 'Securing the U.S. border is a monumental and Herculean task,' said Stein. His series *Border Wars* is striving to present that never-ending, heroic struggle 'at the granular level,' he explained. (Goodwin 2011)

The repetition of the militaristic language and the references to terrorism and weapons of mass destruction legitimate the expensive and aggressive practices at the border.

The Dehumanization of the Immigrants and the Valorization of Border Patrol Agents

The fifth theme that reoccurs in the show is the simultaneous dehumanization of immigrants and valorization of the Border Patrol agents as humanitarian aid providers. These two representations map the good and evil binary that pervades the discourse of the war on terror onto the categories of the Border Patrol agents and the foreign immigrant other. "Border Wars" humanizes the agents through the depiction of casual interactions during the show. The immigrants crossing the border are more elusive and often appear as blurry, pixilated faces, which is ostensibly done to protect their privacy. The resulting visual image, however, goes beyond simply protecting privacy because it simultaneously implies guilt and dehumanizes the immigrant as a faceless other. Eyes and facial expressions are extremely important for eliciting sympathy and for judging intentions. When those are missing, it is much easier to assume all of the people detained are more than likely in the country illegally and possibly a violent threat.

The Border Patrol agents also use dehumanizing language to describe immigrants. For example, segments shot from the Black Hawk helicopter often show immigrants in short, shaky clips as they are running in the desert. Rather than referring to them as people, immigrants, or suspects, the Border Patrol uses the term "bodies." In one episode you hear the pilot say "Ok, we got visual on the bodies." In another, "Everybody is running. We have your bodies." Although there are many deaths in the desert and many bodies recovered, here they are referring to living people, but in a clearly dehumanizing way.

"Border Wars" emphasizes the environmental dangers of the Sonora Desert along the border in Arizona, which result in many dead bodies. The Border Patrol finds about 400 per year despite a decline in total apprehensions. While in 2000 there were 1.6 dead bodies recovered per 10,000 apprehensions, in 2009 the rate climbed to 7.6 dead bodies per 10,000 apprehensions (Haddal 2010).[8] The increase is partially due to a decline in total apprehensions, but the main reasons are the new border fence, the substantial increase in the number of agents patrolling the border, and the changes in enforcement techniques. Easier routes between populated areas are now closed off, forcing immigrants to use longer and more dangerous routes through the desert.

The show, however, ignores the role of the Border Patrol and the border fence in funneling people to these dangerous areas and instead holds up the agents as valiant rescuers that save lives. The first episode of Season 1 describes the section the Sonora Desert near Nogales as a parched, rugged, and dangerous place. The voice-over states: "Some call it the devil's highway because it is littered with the bones of those who thought they saw an easy way into the US and ran afoul of the elements. If it weren't for the search and rescue units like [Agent] McClafferty's, the death toll would be even higher." Every episode of the first season includes footage of the agents providing water and medical care to tired and thirsty immigrants.

The result is that Border Patrol agents, despite having all of the weapons and military gear, are humanized and made out to be caring individuals who are doing their job by helping people. Not only are they bravely facing terrorists and drug smugglers, they are also saving lives of poor unfortunate people that the show emphasizes were abandoned by their wily and unreliable guides. The immigrants, conversely, are either shown in shaky dark images running or handcuffed with their faces blurred. As with many geopolitical narratives, the absences and erasures in "Border Wars" are as significant as what is actually portrayed on the show.

---

8   These numbers refer to bodies found by the Border Patrol. Undoubtedly there are many more deaths that go unreported. The ACLU estimated that there were 5,000 deaths between 1994 and 2009 due to the funnel effect that directs immigrants to more dangerous locations (Jimenez 2009).

Conclusion: Making the Border in a TV Studio

Although at first the war metaphor seems apt for "Border Wars" when compared to other shows like "Storage Wars," "Parking Wars," or "Cupcake Wars," the militaristic language is far more problematic precisely because the border does seem to resemble a war. The viewer knows that wars involving cupcakes are hyperbole and is in on the joke. When the viewer sees the machine guns, Black Hawk helicopters, and Predator Drones patrolling the US–Mexican border it is not a joke at all and the perception that it is a war is strengthened. The confident and authoritative voice-over, the constant repetition of images of war making devices, and the use of fear-inducing hypothetical statements leave the viewer with the strong impression that these extremely aggressive tactics are indeed necessary.

Borders provide a unique challenge for the practice of sovereignty in a territory because just beyond the borderline lies another sovereign state with its own laws and enforcement regimes. Consequently, borders are critical places to impose authority because they represent the first opportunity to identify, classify, and organize the people and things entering the states territory. All contemporary sovereign states have special laws that recognize the importance and challenges of this role and give border agents expanded authority to monitor the area and stop people who could potentially be a threat to the state. The US–Mexico borderlands is a place with a long history of the expansion and (re)-territorialization of US sovereignty through war, the coerced sale of land, the settlement of Anglo populations, the re-signification of the landscape, and now through aggressive and exceptional border enforcement practices. The sovereignty of the state over these lands is not a finalized thing, but rather is reproduced through the daily practices of the Border Patrol and the representation of that space through media. Resistance to the suspension of rights and the militarization of the borderlands is defused through the banalization of war on TV, through reminders that there are wars all around us, and through representations of the border—and the state on the other side—as dangerous, chaotic, and a threat to a civilized way of life. "Border Wars" legitimates the border as a site for the performance of sovereignty and the militarized Border Patrol as a legitimate element of the practice of sovereignty in the state's territory.[9] It also obscures the fact that the border is only one facet of complicated immigration and drug transportation networks. These networks stretch from South America, through Mexico and the border, and end up in the interior of the US. On the show, however, the border is emphasized and the huge market in the US for drugs and the factors that shape immigration decisions are hidden (Heyman 2008).

The images and narratives in "Border Wars" attempt to bring clarity to a range of events, objects, and groups that are located in the borderlands (Castells 2010). The border's complex history of land dispossession, inequality, labor recruiting, and corruption is boiled down to simple language and images. It is a "game of

---

9   Thanks to Joshua Kurz for suggesting these implications.

cat and mouse" where law breakers are brought to justice and even have their lives saved by the valiant US Border Patrol. For viewers, the images of "Border Wars" define Mexico as an uncivilized and violent place: as the show reiterates, "a different world." They define the borderlands as a dangerous place where every trail is a smuggling route, every encounter is potentially dangerous, and everyone in that space is a suspect until they can prove they are not. They create the image of the Border Patrol agent as a patriotic, brave, and compassionate human being who does everything possible to protect "us" from "them." They create the still-blurry image of the immigrant as poor, helpless, gullible, and unsophisticated. They create the image of the smuggler as a "cartel foot soldier" who is armed and willing to do anything to protect their load. The show fills in the details that confirm a series of assumptions about who is doing what along the border. The producers use various cuts, sound and visual effects, and tension building devices to deliver a clear, coherent, thrilling, and, in the end, heartwarming story of the border and the Border Patrol every week.

Nevertheless, the US–Mexico border is not a drama produced in a television studio but rather is a real space inhabited by real people. It is here, at this disjuncture, that the story of "Border Wars" is written. Despite the best effort of the producers, and officials at the Border Patrol itself, to create a clean picture of right and wrong and good and evil at the border, an alternative formulation persistently creeps into this reality show. Although the promotional material, the lead-ins to each episode, and the set-ups for each segment create a sense of foreboding, danger, and imminent threat, the conclusions to each segment never live up to this potential and often paint a completely different picture of who is crossing the border and why. Here we see regular people, families, and long-time residents of the US who simply want to make a better life. We see people who are willing to cross through harsh deserts to go to work because other easier routes were closed by the Border Patrol. We see a woman who went back to Mexico to be with her dying father. We see a family trying to find a better home for their children. We see people who pose no threat at all to the agents, but rather are deeply afraid of the helicopters and guns of the Border Patrol. The desert is not a battleground, the border is not ground zero of a war, there is not a siege of Nogales, and there are not cartel foot soldiers. So far, there have not been any terrorists. What the show depicts is a law enforcement agency with overwhelming and disproportionate military force deployed against a "problem" that could almost certainly be solved in cheaper, more humane ways. In "Border Wars," there are two competing stories of the border; which image remains in the viewer's mind is an open question.

References

Ackleson, J. 2005. Constructing Security on the US–Mexico Border. *Political Geography*, 24(2), 165–84.

Anderson, B. 1991. *Imagined Communities: Reflections on the Origin and Spread of Nationalism*, 2nd Ed. London: Verso.

Andreas, P. 2009. *Border Games: Policing the US–Mexico Divide*. Ithaca: Cornell University Press.

Castells, M. 2010. *The Power of Identity*, 2nd Ed. Malden, MA: Wiley-Blackwell.

Cavanaugh, M. and Heilbrunn, S. 2010. Behind the Scenes of National Geographic's Border Wars Documentary. *These Days*. [Online]. Available at: http://www.kpbs.org/news/2010/jan/05/border-wars/ [accessed: June 5, 2013].

Coleman, M. 2003. The Naming of "Terrorists" and Evil "Outlaws": Geopolitical Placemaking After 11 September. *Geopolitics*, 8(3), 87–104.

Coleman, M. 2005. U.S. Statecraft and the U.S.–Mexico Border as Security/Economy Nexus. *Political Geography*, 24(2), 185–209.

Dodds, K. 2001. Political Geography III: Critical Geopolitics After 10 Years. *Progress in Human Geography*, 25(3), 469–84.

Dunn, T. 2009. *Blockading the Border and Human Rights: The El Paso Operation that Remade Immigration Enforcement*. Austin: University of Texas Press.

El Paso, City of. Website. 2011. Available at: www.elpaso.gov [accessed: June 5, 2013].

Fairclough, N. 1995. *Critical Discourse Analysis*. London: Longman.

Finklea, K. 2013. *Southwest Border Violence: Issues in Identifying and Measuring Spillover Violence*. Washington, DC: Congressional Research Service.

Foucault, M. 1971. *The Order of Things: An Archeology of the Modern Sciences*. New York: Random House.

Foucault, M. 2002. *Archeology of Knowledge*. New York: Routledge.

Goodwin, J. 2011. GovSec 2011: Nicholas Stein, producer of "Border Wars," Describes a Herculean task. GSN Magazine. Available at: http://www.gsnmagazine.com/node/22855.

Goffman, E. 1979. *Gender Advertisements*. New York: Macmillan.

Gregory, D. 2011. The Everywhere War. *The Geographical Journal*, 177(3), 238–50.

Haddal, C. 2010. *Border Security: The Role of the Border Patrol*. Washington, DC: Congressional Research Services.

Haddal, C., Kim, Y., and Garcia, M. 2010. *Border Security: Barriers along the US International Border*. Washington, DC: Congressional Research Services.

Heyman, J. 1998. State Effects on Labor Exploitation: the Ins and Undocumented Immigrants at the Mexico–United States Border. *Critique of Anthropology*, 18(2), 157–80.

Heyman, J. 1999. Why Interdiction? Immigration Law Enforcement at the United States–Mexico Border. *Regional Studies*, 33(7), 619–30.

Heyman, J. 2008. Constructing a Virtual Wall: Race and Citizenship in U.S.–Mexico Border Policing. *Journal of the Southwest*, 50(3), 305–34.

Heyman, J. 2009. Trust, Privilege, and Discretion in the Governance of the US Borderlands with Mexico. *Canadian Journal of Law and Society*, 24(3), 367–90.

Heyman, J. and Ackleson, J. 2009. United States Border Security after 9/11. *Border Security in the al-Qaeda Era*, edited by Winterdyk, J. and Sundberg, K. New York: CRC Press, 37–75.

Jimenez, M. 2009. Humanitarian Crisis: Migrant Deaths at the US–Mexico Border. San Diego: American Civil Liberties Union.

Jones, R. 2009. Categories, Borders, and Boundaries. *Progress in Human Geography*, 33(2), 174–89.

Jones, R. 2012. *Border Walls: Security and the War on Terror in the United States, India, and Israel*. London: Zed Books.

Lytle Hernandez, K. 2010. *Migra! A History of the US Border Patrol*. Berkeley: University of California Press.

Marosi, R. 2011. April 21. Plunge in Border Crossings Leave Agents Fighting Boredom. *Los Angeles Times*. [Online]. Available at: http://articles.latimes.com/2011/apr/21/local/la-me-border-boredom-20110421 [accessed: June 5, 2013].

Müller, M. 2010. Doing Discourse Analysis in Critical Geopolitics. *L'Espace Politique*, 12, 1–21.

Nevins, J. 2010. *Operation Gatekeeper and Beyond: The War on "Illegals" and the Remaking of the US–Mexico Boundary*. New York: Routledge.

Ó Tuathail, G. 1996. *Critical Geopolitics: The Politics of Writing Global Space*. Minneapolis: University of Minnesota Press.

Ó Tuathail, G. 2002. Theorizing Practical Geopolitical Reasoning: the Case of the United States Response to the War in Bosnia. *Political Geography*, 21(5), 601–28.

US Customs and Border Protection. 2004. Frontline Defense Against Terrorism is Priority Mission as US Customs and Border Protection Commemorates One-year Anniversary. March 1. [Online]. Available at: http://cbp.gov/archived/xp/cgov/newsroom/news_releases/archives/2004_press_releases/032004/03022004.xml.html [accessed: June 5, 2013].

US Customs and Border Protection. 2005. National Border Patrol Strategy. Washington, DC: US Border Patrol.

Chapter 10

Latin American Borders on the Lookout: Recreating Borders through Art in the Mercosul

Anne-Laure Amilhat Szary

Do Latin American borders exist? This provocative question can be useful to make my point and introduce the argument of this text. Of course, Latin American countries emerged as modern nation states endowed with borders, which began to be traced as international limits, first with the separation of the two Iberian empires and then with the nineteenth-century decolonization process. Remarkably, however, territory does not play the same role in the "Extreme Occident" (Rouquié 1987), which this part of the Americas represents, as in the old Western world. Indeed, if territory building chronologically represents the first step of stato-genesis (nation-building only came thereafter, once the borders had been drawn on maps), after a very nationalist nineteenth century, borderlands were soon evacuated of spatial imaginaries. This is partly due to the fact that they lay in scarcely populated regions where the detail of their demarcation was of no essential interest to the capital and they rather appeared as frontiers, full of opportunities (Perrier Bruslé 2007). We will come back to the legacy of this regional history which does not exclude violence and war in the process of boundary-making, but which conditions border narratives, offering Latin American borders an exceptional status.

It is never simple to pretend to examine a continent's borders as if they possessed unified characteristics. This point is underscored by the case of Latin America, which is separated from its northern counterpart not by an isthmus but rather by the Mexico–US border. "Latin America" is born of the encounter of a political project based on strong cultural power relations and can be described as a strong "idea" (Mignolo 1991) that is rooted largely in alterity. The dominant discourse has long been a credo in "mestizaje"/"mestiçagem" (Gruzinski 1999), a term resembling the US term "melting pot" since it was, in Spanish or Portuguese, explicitly about interbreeding and the mixture of bloods. Paradoxically, as hybridity has become a new paradigm to conceptualize globalization, "mestizaje" lost political ground since the quincentenial of Columbus' voyage. The post-1992 historians strived to show how racism led to a long-lasting political and social exclusion in the region, through a process of domination which was for centuries hidden by the belief in the virtues of intermixture (of plants, techniques, etc.). If we define borders as places where alterity has to be confronted (Amilhat

Szary 2012a), their understanding therefore highly depends of the segregation versus integration processes that play out where a border works. The analytical context of contemporary Latin American borders must therefore be postcoloniality, while not overlooking ongoing global processes that continue to shape them. Postcoloniality allows us to link global domination processes with the challenging attitudes of Latin American elites towards the descendants of indigenous groups that persist to this day on the continent. Borders in Latin America have long been shaped by both local politics and external powers that regularly brought in arms, war alliances, conflict resolution arbitration, and all the ideology to sustain these.

What I offer here is a focus on contemporary local artists and art promoters who put borders at the core of their endeavors over the last decade. I also suggest that this outburst of creativity could be seen as a precursory sign of a switch in regional border narratives. Creative actors cannot be considered to be representative of their fellow citizens, but they have the power to express their representations in a way that possesses a strong power of conveyance. I will discuss here the emergence of "border art" and analyze it according a methodology elaborated through research in other continents (Amilhat Szary 2012b, 2012c, 2013). My epistemological guiding line lies in the performing power of visual arts on the border. The artwork not only reflects a spatial reality: it operates in both the representational and non-representational (Thrift 2008) modes. The artists design the place that they work with, in a totally relational process, through the mobilization of aesthetic power. As a sensitive object, the artwork not only conveys the artist's explicit representation to the audience, but is also liable to provoke original feelings within the person who becomes acquainted with it. This aesthetic dialectic makes them both co-producers of the border reality. It is therefore not very pertinent to analyze art production in conflictive places only as resistance processes: even when denouncing a situation, they mediate it in a way that partly contributes to its reproduction. It is not possible to consider this relation between Latin American borders and visual arts exhaustively, either in terms of covering all of the dyads or regarding the national origin of artists. I have chosen to focus on the emergence of border concerns and art projects at the heart of the continent, analyzing how the Mercosul Biennial—an art fair that began in 1996 in Porto Alegre, Brazil—has contributed to this process. The next section provides additional context on the status of borders in Latin American spatial imaginaries, and it is followed by an examination of the recent emergence of borders within the Bienal do Mercosul exhibitions, an art event which has accompanied the restructuring of the region around the claim for centrality on behalf of Porto Alegre. The final section analyzes Marina Camargo's work, a young Brazilian artist who displayed in 2011 an installation entitled "*Tratado de Limites*" (Boundary Treaty).

Bringing Borders Back into Latin American Spatial Imaginaries

When considering Latin American territorial issues, resource management or urban inequalities usually first come to mind. The Latin American region offers an example of what appear to be non-conflictive and non-problematic borders: during the twentieth century, in most cases, for the Latin American armed forces, the enemy lay within national borders. This obviously has to be nuanced. In the Latin American case, two factors combine to present a border panorama which is indeed quite homogeneous, if not as peaceful as imagined. First, political limits were drawn primarily in areas with very low population densities. Second, they were designed in the course of a process which implied border delineation from a great distance in time and space, which led Michel Foucher to describe them as "meta-borders" (Foucher 2012 [2007]). Both factors are the result of the colonial process to which the continent was subjected beginning with the arrival of Europeans in the Americas (Gruzinski 2004).

The two empires that competed to establish their rule did not need to immediately fight for the land between them: upon the return of Columbus to Europe, Spain and Portugal resorted to Papal arbitration to set up a just sharing process of the "new world." Pope Alexander VI signed, along with representatives from both countries, a treaty in 1493 in the small town of Tordesillas that split the space of conquest in two. In a very modern way, the supreme pontiff resorted to tracing a line on a map, vertically, 100 miles from the western-most position known at the time. Portuguese possessions were to expand east of the line, Spanish ones over its other side. The existence of such a geometric division of sovereignty is much more significant than the location of the limit (which moved various times as the actual size of the new continent was determined and the Portuguese did their best to enlarge their share of land). The term "meta-border" was inspired by the fact that a border was drawn before the lands it divided were known. In a sense, a great majority of subsequent colonial border making also resulted in meta-borders, since they were designed without field demarcation and according to international and newly emerged national power balances that did not take local forces into account. The period of Latin American independences in the eighteenth and nineteenth centuries contributed to the fact that Latin America could be described as a continent of formal limits. Indeed, an international principle was then adopted (at the Angostura Congress of 1819, held under the presidency of Bolivar) that later took the name of its first Latin words, *Uti Possidetis*, that required the international borders of the newly created states to respect previous colonial limits.

Common interpretations of the invention of Latin American borders argue that they emerged in a context of great territorial stability. However, the principle of *Uti possidetis* was not followed as borders were being drawn, and upon closer examination one realizes that Latin American borders are not the direct legacy of division of the continental space created by the Iberian empires (the latter having pushed their frontiers as much as they could, in a territorial rivalry which was very sensitive in the "gaucho" area where Porto Alegre lies). While the capitals of new Latin American states were often the same old centers of colonial power, this does not hold true for their peripheries. A detailed analysis by Foucher (1991) shows that, in fact, only 30 percent of international borders followed lines existing prior to the nineteenth century, while 10 percent were delimited after the opening of the

Panama Canal in 1914. This implies that the vast majority of them (60 percent) were defined between 1800 and 1914; that is to say after the independences and as the result of adjustments made by the new states rather than through a colonial legacy. This was due to both a lack of precision in colonial limits, which had seldom been formally demarcated, and to their multiplicity (viceroyalties, audiences, general captaincies, provinces, and *gobernaciones* or *comarcas*). The determination of the states' limits of the ex-Spanish Empire territories and Brazil was therefore based on a pragmatic adaptation of the principles described above, known as *de facto* "*uti possidetis*." The negotiation was based on the land by then occupied by the Portuguese, who had greatly expanded their area of influence in the Amazon basin, and through conflict between the new states. One could claim that nationalism was more completely built in South America than in Europe, because of the erasing of previous cultures. Thus Latin American border-making was not as peaceful as often claimed. The region of focus in this chapter is indeed the place where the Spanish and Portuguese crowns met and fought, over the possibility of the existence of Uruguay as an independent country. However, many conflicts did not turn into wars due to a common practice of resorting to external mediation. The continent's borders can be described as the product of a marginal remaking of colonial contents, sometimes exacerbating it and sometimes devising transitional agreements.

Once established, there were few significant disputes at Latin American borders with the exception of irredentist claims such as Bolivia's regarding access to the sea lost to Chile. This lack of conflict is the result of the borders' locations: most passed through sparsely inhabited areas where the people who actually came into in contact with them have long been forgotten in nationalistic historiographies. From central regions of the new states, border areas were treated as peripheries with potentially abundant resources. In the case of the Pampas, both Urugayan and Brazilian authorities had active borderlands policies, sending new immigrants to establish localities on both side of the border. Moreover, a tradition of great mobility existed among the descendants of the pre-Hispanic populations, at various scales.

Recent scholarship has identified other processes such as regional integration narratives progressing together with rebordering issues at the Latin American borders (Amilhat Szary 2007). The security turn, which shaped much of global politics since 2001, contributed to the globalization of border management issues (Brunet-Jailly 2007) and resulted in a new emphasis on borders in Latin America. The simultaneous launching of two border programs in Brazil and Chile involving high-technology is symptomatic of the securitization of borders, although the two programs operate at different scales. Brazil launched a massive project in July 2011 called SISFRON (System for the Surveillance of the Frontier)[1] at a

---

1   SISFRON is intended to complement the already existing System for the Surveillance of the Blue Amazon (SISGAAZ), a coastal surveillance program from the Brazilian Navy. These two new programs would build on the SIVAM (System for Vigilance of Amazonia) program established nine years before by the Brazilian Air Force and Waltham (Raytheon Co.).

total cost estimated to be on the order of US$6.3 billion by 2019. In October of the same year, Chile initiated the *Plan Frontera Norte*, a regional scheme slated to cost $70 million through 2014. Both are technology driven policies based on satellite surveillance, new planes, unmanned aerial vehicles, and unattended ground sensors, all of which are intended to add additional layers of coverage with few new human resources. They relay information from the ground to hasten the building of integrated border facilities, to render goods monitoring and people's crossings more easily tracked through biometrics and tagging. These are ambitious goals and elusive proposals given the relatively low number of border crossing points (31 for the whole of Brazilian territory) and the length of the borders. In this context, the "openness" of Brazilian borders has rapidly switched connotations.

In Brazil, a dedicated Commission at the Congress (*Subcomissão Especial para Acompanhar as Ações de Proteção* às *Fronteiras*), within the Foreign Affairs and Defense Commission, was launched in 2011 to promote the rebordering agenda which resulted the same year in the promulgation of a "National Strategy for the Borders," ENAFRON, by the Brazilian Government. This move formalizes the government's rising interest in border control which had previously led the National Integration Ministry to conduct a full diagnosis of its border lands (Brasil, Ministério da Integração Nacional, 2005). This project was run by RETIS, a team of geographers under the leadership of Lia Osorio Machado at the Geography Department of the Rio de Janeiro Federal University (http://www.retis.igeo.ufrj. br). RETIS has successfully developed networks throughout the country, notably around Adriana Dorfman's work on smuggling at the Rio Grande do Sul University. The rising number of publications from this team attests to the emergence of a scientific approach to borders in Latin America, while the growing budgets signify that public opinion favors investing in a relatively new border monitoring project.

Borders, a Rising Issue at the *Bienal do Mercosul*[2]

This section analyzes the recent evolution of border imaginaries through the emergence of this territorial dynamic at the Mercosul Biennial, a visual arts show held in Porto Alegre, Brazil. The first Mercosul Biennial, organized in 1996, five years after the official creation of the South American common market, and the year following its implementation, was a formidable challenge. São Paulo had already held its own biennial, which meant that a second Brazilian contemporary art event had to position itself in light of the first. Their common point was their location in the country's most economically powerful cities, not in the political capital of Brasília, thus inscribing art within a valorization process of global cities, along the lines of Florida's work on creative cities (Florida 2002). More than just indicating that art is a powerful market that intersects with other sectors,

---

2   According to the language they use, the spelling of this event name varies: *Biennial* in English, *Bienal do Mercosul* in Portuguese, *Bienal del Mercosur* in Spanish.

the biennials illustrate how local leaders can promote cultural action in order to promote their cities.

The eighth biennial art fair was held in Porto Alegre in 2013, underlining the success of the initial dual purpose bid of 1996. Hosting a contemporary art fair meant that Porto Alegre was positioning itself as a global city, fighting against mainstream rankings such as Peter Taylor's Globalization and World Cities Group at Loughborough University first ranking of world cities only included a handful of Latin American metropolises (Beaverstock, Smith, and Taylor 1999). The name also attempted at grab a central position within the emerging regional bloc of Mercosul (Mercosur in Spanish) since Porto Alegre was so close to the Argentinean Border. "In a laudable strike against regional chauvinism, they decided to take the southern cone of South America as their cultural area, taking the name of the recently establish trading block, Mercosul, as their frame of reference," wrote Gabriel Pérez-Barreiro in the tenth-anniversary catalog (Morgan, Paternosto, and Pérez-Barreiro 2007: 182). The chief curator of that sixth Biennial wrote: "While this term was not without its problems, considering the failure of Mercosul to create effective regional integration, and the invisibility of this term outside South America, it did set an area for action that was both regional and international" (Pérez-Barreiro in Morgan et al. 2007: 182). By highlighting the poorly considered balance of the free-market zone, the curator stresses the much greater reach of the Biennial. It may not be completely accurate to state that the Mercosul treaty was not reaching its goals after 10 years of existence, since it had by then succeeded in stabilizing Argentine–Brazil relations, produced a series of bilateral agreements, notably with the European Union, and increased both the internal (economic interdependence) and external trade of the bloc. Although a Parliament had just been created in 2005 at the time of the sixth Biennial (tenth-anniversary edition), the Mercosul common market bloc was perceived as a distant, economically oriented construction, with very little impact on citizens' lives.

Taking contemporary art from one Brazilian economic capital to another entailed reorienting the perspective on the displayed art. If São Paulo remained the country's conduit to international art markets, presenting works that were interconnected to the vast network of global contemporary production, the art event in Porto Alegre needed to express a different relation to space. Describing the contents of his exhibit, curator Pérez-Barreiro wrote, coming back to the original objectives of the Porto Alegre project: "The curator of the 1st Mercosur Biennial, Frederico Morais, took the challenge to organize and structure an alternative and non-Eurocentric history of Latin American art, describing ideological vectors into which art from different Latin American countries were placed" (Pérez-Barreiro in Morgan et al. 2007: 182). Bringing contemporary art out of São Paulo certainly made an artistic statement. However, once the first Mercosur Biennal was finished, the following editions were disappointing to a geographer's eye, since they gradually left aside this project of locating aesthetic production within a global/ local perspective, drifting towards more classic exhibition choices, where mainly Brazilian artists were presented together with those of some Mercosul countries.

Notwithstanding the connections that could have been made between the art and the place where they were displayed, it was clear that "Porto Alegre today has probably had more exposure to Latin American art from more countries than any other city in Latin America" (Pérez-Barreiro, in Morgan et al. 2007: 183). Cultural components such as history or language were at the forefront of the curatorial effort, together with the gradual opening to non-Mercosul, and even non-Latin American artists.

After 10 years of existence, the 2005 Porto Alegre Biennial was facing some organizational issues which required a redefinition of its purpose. This meant making its multi-scalar spatial relations more explicit:

> While the leaders of the biennial were keen to expand the geographic range of the biennial and further internationalize it, the context for this decision was the saturation of international biennials all over the world over the last decade, and the relative similarity among them. The pre-existing regional structure of the biennial had simultaneously limited its range and protected it from becoming another stop on the international contemporary art circuit. (Pérez-Barreiro in Morgan et al. 2007: 183)

A decisive curatorial choice was made in order to provide of an "intermediate model" that would not rely on the duality of "national representation on one hand or global consensus on the other." The decision taken by the board was to head for a multi-curator event that could expand to a greater number of locations in the city. This also constituted a potential opening for a larger and more diverse audience, since "This question […] implicitly contained the issue of who the biennial should serve, a local or international audience? By truly serving a local audience, could the biennial establish a new model or paradigm that would thus also distinguish it in the international arena?" (Pérez-Barreiro in Morgan et al. 2007: 183). The goal was achieved by hiring for the first time a non-Brazilian head curator who was tasked with deconstructing the link between art and space that the first Mercosul had begun to establish in favor of something more original. Gabriel Pérez-Barreiro suggested a local focus to tackle the postmodern figure of in-betweenness. By drawing on the metaphor of the "third bank of the river," the title of a famous short story by Brazilian author João Guimarães Rosa,[3] he suggested "a radically independent space, a space free of dogma or imposition, a place of observation" (Pérez-Barreiro in Morgan et al. 2007: 183):

> What if the artist rather than the national curator or the diplomat were allowed to draw the map? The challenge would be to find a model that was still rooted in the Mercosur region, but that was not limited by it. The solution was to discuss a biennial from rather than of Mercosur. (Una bienal desde, a partir del Mercosur) (ibid.)

---

3   From his 1962 collection of fantastic short stories, *Primeiras Estórias*.

This introduction of relational thinking within the contemporary art project allowed him to split the Biennial into various interdependent projects bearing very geographic titles. The Conversa exhibition, for example, allowed Mercosul artists to invite fellow creators, based on thematic and creative affinities rather than places of origin, whereas the Zona Franca section relied on curators' choices. "[They] had to be freed from their traditional role of representing their countries, and be given the freedom to work on each project without limiting the geographical range of their choices" (Pérez-Barreiro in Morgan et al. 2007: 184), with a very specific mission, that of initiating *in situ* creation.

It was not by coincidence that the place chosen to bring the invited artists to propose a new work for the sixth Biennial in 2007 was the *Três Fronteiras*, the tripartite border zone between Brazil, Argentina, and Paraguay, a site that symbolized Mercosul's diversity and complexity. Ticio Escobar, who was in charge of this session, offered four artists a residence in the region: three from Latin America (Minerva Cuevas from Mexico, Jaime Gili from Venezuela, Aníbal Lopez or "a-1 53167,"[4] from Guatemala) and a Bulgarian, Daniel Bozhkov. Interestingly, Bozhkov's proposal incorporated more of the atmosphere of the region as imagined from a European point of view, by introducing Guarani basketry work into a contemporary art display, whereas the three others let themselves be inspired by the Tripartite Point, where there is a cross-border region of three linked cities, Ciudad del Este, Paraguay, Puerto Iguazú, Argentina and Foz do Iguaçu, Brazil. Even if many tripartite borders are geopolitically significant in Latin America, none is as well-known as this one, generally referred to as "LA triple fonteira." The interpretation of the tripoint given by Brazilian artist, Beto Shwafaty, called "Tripartite Reunited" (2011), consists of a sculpture based on the Möbius strip and its geometrical dynamics.[5] According to him, it reflected the merging of artistic and political preoccupations in such exceptional places. It is indeed a place of unique symbolic power: "selva" for the Argentinian, "frontier" for the Brazilian, resource and gateway for the Guaranis and Paraguayans, and entry port to globalization and enrichment for the many communities who live there, be they Palestinian, Chinese, or Korean.

During the sixth Biennial, only one of the guest artists produced a work that directly tackled border issues, even within the *Três Fronteiras* section. A. Lopez's unnamed proposal is referred to as "The Guardian" in the exhibition catalogue. His piece was inspired by the smuggling industry that takes place on the Parana River border. Trafficking of kinds of goods is proliferating, much of it in plain sight, involving a very delineated, almost choreographed, game of cat and mouse between couriers and authorities. The "sacoleiros" throw boxes into the river on one bank while others retrieve them on the other, having to undergo a number

---

4 He generally uses his ID-Number to sign his works.

5 http://www.shwafaty.org/?portfolio=tripartite-reunited. He also refers to Max Bill's work called "Unidade Tripartida," which is a very important reference to Brazilian, currently being shown at Brazilian Pavillion, in Venice Biennale (autumn 2013).

of negotiations—including bribery—with military and customs representatives in order to be able to get away from the border zone. The local benefits of this exchange system lie in contrast with the imposition of stricter border management in the area, and the growing harassment of smugglers is the result (see Miggelbrink, this volume). Fascinated with the hypocrisy of the situation and by its theatricality, "a deeply rooted culture of smuggling that colors an array of practices and social representations" (Escobar and Pérez-Barreiro 2007: 159), "a-1 53167" decided to smuggle his work through:

> Bringing this vast system to bear on art demands a strategy for approaching and delving into this issue through subtle, precise and energetic moves: a rhetorical movement capable of establishing a distance from it. But that distance must be minimal; by moving too far away, the artist would lose the power that the situation unleashes. (Escobar and Pérez-Barreiro 2007: 159)

He planned the packing up of empty cardboard boxes to be covered by black plastic bags and allowing them to cross the river from Paraguay to Brazil, then their subsequent transfer to Porto Alegre where they were piled up in the Biennial exhibition hall. Once again, if the artwork resists the system, it is not only by showing and sharing the representation of a place but by the capacity that the creator develops to perform it. Here, "a-1 53167" displaces the flows and control issues of the border contained within the apparently empty boxes from the Tripartite zone to Porto Alegre to convey it to the Biennial visitors.

It took two more Biennials to see the theme of borders overlap and become central to the whole event. The eighth Biennial, which took place in 2011, was entitled "Essays in Geopoetics." Its objective was not only to provide a temporary artistic display but also to set the basis for a locally grounded initiative to foster spatial awareness. According to the 2011 chief curator José Roca:

> In terms of the choice of theme, I tend to think of a theme that can also be a strategy for curatorial action. In my work as a curator I have tried to reconsider the model of the Biennial so that it can be adapted to local conditions and more effectively reflect the local scene where it is taking place. (Roca in Dias Ramos and Roca 2011: 15)

It is very interesting that it should be within this curator's system that many artists tackled the issue of borders, considered "not just as a 'thematic framework' but as a form of 'activation strategy'" (Roca in Dias Ramos and Roca 2011, ibid.) liable to diversify the relation to place of the visitors and inhabitants of Porto Alegre in general. The art works were chosen so that performing the border could happen in a very subtle interaction with the international line. The context of the eighth Biennial was generally that of a more complex relationship to space, which includes relations to both here and there, and it mixed place experience with multiscalar imaginaries. The curator himself referred to the "notions of locality, territory, mapping and border" (Roca in Dias Ramos and Roca 2011: 16). Sections of the art fair carried such evocative names as "Travel notebook" (*Cadernos de Viagem*), "Continents," and "Beyond borders" (*Além Fronteiras*). It was in the latter that border standpoints were most developed, since the show was built in order to share a critical view of the Rio Grande do Sul landscape through the work of nine invited artists, among which I have chosen to detail *Tratado de Limites* ("Boundary Treaty"), the work of a young Brazilian artist, Marina Camargo, in the next section of this chapter.

Marina Camargo's *Tratado de Limites*

The eighth Porto Alegre Biennial, which opened its doors in September 2011, was focused on the local-national dialectic and border issues: they appeared as a paroxysmal expression of complex relations to place. This came with more intensity in the section curated by Aracy Amaral, "Beyond borders" (*Além Fronteiras*), where *in situ* work was commissioned to local artists of different generations and it is interesting to consider this section alongside the perspective that was taken in the main show. The principal section of the eighth Porto Alegre Biennial, under the title of *Geopoetics*, presented works dealing with land and space, one of which can be highlighted for its capacity to tie the local and global as well as questioning the borders of identity from a Brazilian point of view. It is called *Bandeiras* and signed by one of the most famous contemporary visual artists, Emmanuel Nassar. It is interesting to use *Bandeiras* as an introduction to Marina Camargo's *Tratado de Limites* because it stresses the power of national imaginaries in the "continent-country" that Brazil composes, together with the deep anchorage in localities which the size of the nation also imposes. Citizens are easily captured by this local/national dialectic, now opened to global influences, without really having to experience the national borders which are recalled upon only as symbolic guardians of their territory.

Bandeiras consists of a very large display of flags (390 x 540 centimeters), on three walls that encompasses the visitor and gives him/her an impression of great exoticism, which contrasts with the fact that his/her rational capacities are informing him/her that he/she is probably in front of very banal objects. Nasser drew inspiration from an anthropological exhibition on Ghanaian flags that he visited in Germany in 1993 and *Bandeiras* had travelled quite a lot before, presented for the first time in 1998 in Sao Paulo (see also "Como Fiz Bandeiras," http://artenassar.blogspot.fr). The strength of his contribution is that the 143 colorful pieces that he assembled only present an appearance of territorial identity. They can be immediately identified to flags, the utmost symbol of national belonging, but it is impossible to tag the flags to the country they should refer to because the pieces are not actually country flags. The work presents communal flags from the municipalities of Para, Northern Brazil, the artist's home region. It was a long and participatory process for him to collect them, since the flags do not normally circulate outside of their communities. Nassar had to launch a research campaign in a local newspaper, *O Liberal*, with the slogan "I want your flag" to be able to account for the diversity of his state. He then assembled them without tagging the names of the places they came from. One could think that his technique of "collage" refers to a traditional representation of hybrid identities in a "mestizo" continent, but he likes to deconstruct this. According to the artist, the flags were also chosen for their colorful diversity, in a pop-art perspective where they become objects of consumption that are denounced by his technique. Only one state of Brazil thus appears as a world of its own, giving place-belonging a fractal dimension that the artist intended to apply at the scale of the country or the global level. The 2011 edition of the work, however, presents a great difference to the original installation, since Nassar does not present the flags themselves but a reproduction of them on painted recycled metal plates. The use of old metal[6] to express the identity of a fast developing country is intentionally paradoxical since it materially stresses the "ready-made" component of nationhood that globalization confronts.

The challenge of the section of the eighth Mercosul Biennial in 2011 called "Beyond borders" (*Além Fronteiras*) resided in the capacity of guest artists to work upon their interpretation of the sense of place which the Rio Grande do Sul provoked. The curator, Aracy Amaral, was able to invite a small group of international artists to the region that hosted the Biennial, offering them good conditions for the creation of an original proposal. Marina Camargo was thus able to travel to the border region three times, first with two fellow invitees, Lucia Koch and Cao Guimaraes, and the third time with Biennial staff, who helped her with the logistics of her *in situ* performance. This was even more valued by her since she was then living in Germany for a long-term residency and thus experiencing the Rio Grande do Sul through immersion (a discrepancy which she sees as meaningful, not only because of the material distance it implied). Camargo's idea of borders is not apparently very explicit, either in the name she gives to her exhibition or in the other documents produced, since she proposes to overcome barriers and scales. However, Camargo's installation sparked a lot of debate and interest. Not only was her title directly tackling the border issue, but she was also offering a sound and visual device which included photos and video, and even a strange map that visitors were invited to take home as a token of the work.

The *Tratado de Limites* was based on the documentation of the artist's immersion in the Pampas region, the vast plains which join the South of Brazil to Argentina, via most of the Uruguayan territory. Born in Maceió (Alagoas,

---

6   This is one of his favorite materials, which he had used when featured in a previous exhibition bearing the same name, "Fronteiras," back in 1997 in his native state of Para (it constituted in the 16th edition of the Arte Para art show). Although most of the works of the show only alluded to the geopolitical component of the title, he had struck it frontally, presenting an installation called *Brasil*, including a strange map of his country, *O Mapa em Negro* (1996), featuring the shape of the country in a small red contour painted on a large rusty black metal plate.

Figure 10.1　Marina Camargo, *Tratado de Limites*,
　　　　　　　overview of the installation
Source: Photo by Fabio del Re, with the permission of the author.

northeast Brazil), Camargo moved to Porto Alegre at the age of nine. She began university there, before completing her arts education in Barcelona, New York, and Munich. Although she had travelled the world, working at the Southern border was also her first opportunity to visit one of her own country's borders. She justifies her decision regarding the location of her piece for the Biennial on the basis of a very personal relationship to this place, both affective and interpretative of what she imagines of international limits: "I chose Pampas, because of special relations for my family, and also of its very vast landscape. I was very impressed on how geography could change the political and the borders: frontiers in this area made no sense, because of the isolation of these areas."[7] Later in the interview, she told me that she made this choice on the basis of previous visits, which had never been exhaustive ("I didn't cross the area before, I had never crossed them"). The borders located in the Pampa region offered a contradictory appearance: vast plains contrasting with, on some part of their delineation, "rivers which make borders," and according to her, "this is the kind of borders that make sense, not political borders." The encounter provoked by the Biennial curator represented a

---

　　7　Personal interview with the artist, "skype" voice and camera conversation, September 16, 2013. All quotes from this section of the chapter are extracted from the verbatim of this conversation.

very important turn in her life, both emotionally and professionally: "that work was very special for me: I initiated a different process of work, including more documentation, spending more time on a project." Its spatial dimensions, between living an enriching professional life abroad in Germany and immersing herself into the Pampas borders, have considerably transformed her artistic practices.

This "density of the research" explains the composition of the work finally exhibited, including its title. In the process of documenting the border area she was about to cross, Camargo visited the *Instituto histórico e geográfico do Rio Grande do Sul*, the regional museum of the Historic and Geographic Society of Porto Alegre, and found a copy of the treaties that had established the state contours and the international limits of the countries. The opposition of these dry but powerful words with the openness of the landscape was the founding inspiration for her creation. She thus decided to look for a way to convey this complex experience of place through multiple mediums. These included diverted pieces of information (maps, official texts), pictures of what she discovered, and a soundtrack of the wind in the high prairies. These aspects were connected with a model of the border carved in ice that she photographed while it melted and a separate documentation of an *in situ* performance in Tacuarembó. This plurivocality appears essential to understand her intervention.

She appeals to the relation between what she expected from the border, the upsurges of an apparently very uniform landscape, and the power of human convention to create diversity and identity in an unforeseen process. Two elements of her installation may appear as the most simple to understand for the visitor: two almost black and white photos which represent a rain curtain over the plains. They were taken near Bagé, where the climate progressively gets very arid—almost desert—and where Marina Camargo was surprised to discover another "kind of invisible frontier, that of climate; there, for some reason, the rain can't go further, causing the dryness." They directly answer the "map" that was piled up on the floor in order to be taken away by visitors. She uses picture of clouds in the sky, whose shape could be compared to that of continents on a *mapa mundi*, and she gives life to her oneiric divagation by superimposing a grid of supposed meridians and parallels together with place names as seen in a mirror. The skies reflect our earth and make our dreams of territorial horizons possible. Maps were a very central element of Marina Camargo's work before her participation to the Mercosul Biennial, linked to her first long sojourn abroad. It is interesting to see here how they are deconstructed: from the Atlas and old maps all that remains is only the general title for her work and the take-away document—that, according to her words, should work like a tourism office leaflet—and has only the appearance of a classic, geospatially referenced, representation.

These samples of the region are the most figurative part of her composition. The other very geographic element of her work is more conceptual. It consists of the very condensed documentation of a performance set on the outskirts of a small town of the Uruguayan part of the Pampas called Tacuarembó. She chose Tacuarembó because the small locality has designated itself as the center of this vast region.

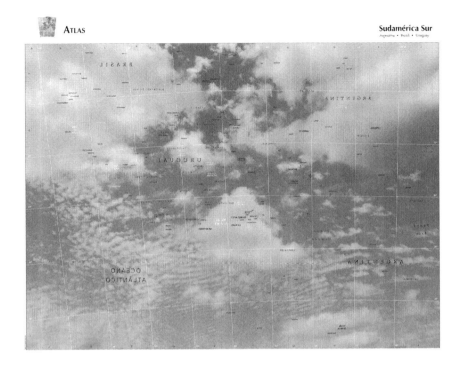

Figure 10.2   Marina Camargo, *Tratado de Limites*, 2011,
atlas map distributed to the visitors

Note: Copyright M. Camargo, with the permission of the author.

This quest for centrality in such a landscape may appear as absurd as the border demarcation that straddles it. More precisely, this little town has reformulated its cultural identity as "capital of the Patria Gaucha:" this process bases itself in the rediscovery of the value of regional belonging when the plains enter the globalized world, which imposed a new intensity to the agri-business in the 1990s. Whereas "Pampas" is a toponym inherited from the autochthonous languages ("plains" in Quechua), the Gaucho identity bears the weight of the colonial frontier. "Gauchos" were partly nomadic Indians described by Europeans as vagabonds, before being used to forge an Argentinian national-type shared with the Brazilian state of Rio Grande do Sul. The meaning has mutated as gaucho was co-opted by the nation and forced into national imaginaries (Ludmer 2002). Camargo decided to make a replica of the city entrance signs, which she loved to find as she travelled. The material shape was not local but referred to another, world-famous, place name on a hill, that of Hollywood. "There is irony to put those letters in a place that is not important to the world, like Hollywood:" this was the kind of collapsing that globalization could allow. The micro-level settings of a zoo, even reinforced the mimicking effect that was intended. She chose the town of Tacuarembó because of its claim to a central position within the region, but she also worked hard to ensure the letters were placed in very precise locations. She insisted on this: "I was also very concerned on how to put this sign there. It was important that it should be buried: maybe it can disappear into the landscape, it can melt down into the landscape when the grass grows" (the process of melting also connects with her ice carving). She therefore envisioned burying the letters that were to be displayed at the entrance to the urban area so that they would little-by-little be incorporated into the vegetative landscape that so fascinated her. But her intervention was looking for a way to interrupt the evenness of the Pampas. In her own way and words, she was "making some point on the landscape" and, in turn, using the power of demarcation. This was undoubtedly tied to her fascination for maps, and her "concern about how to mark a position in geography."

The intensity of the performance lies in its power to provoke significant events. In this case, it proved very difficult to bring the letters from Brazil to Uruguay. "But because of this, the work almost did not happen because of the bureaucracy. [...] This is very important information to me: I was somehow provoking the idea of the Biennial title, I was dealing of the limitations of the treaty." Although the art show was intended to promote the Mercosul regional integration process, its field implementation was bluntly denied. A personal incident also allowed her to measure both the permeability and controlling issues at stake on the border. She recalls, "When I went to Uruguay, I presented my passport, but I didn't have to do so upon going out, because it was too easy, so when I came back, I had to pay a tax for leaving the country illegally." Borders were resurfacing at unexpected locations as collective identity building disrupted the personal experience of landscape. That could not only be conveyed through images, which is why she included the soundtrack (by Leonardo Boff) that would allow the spectator to feel the wind and its waves, both in the air and in the high weeds.

The most visually astounding part of Marina Camargo's *Tratado de Limites*, and the one which represents the biggest element of it, is the carved piece of ice that is a model of a three dimensional map of the Plata region where the Pampas rivers flow into the ocean. After sculpting it, she let it melt while regularly photographing it in order to make an animation film, "Geografia (Paisagem com Ondas)" (http://vimeo.com/51392489). The fact that she did not video it also reveals her manner of tackling the relation between trace and process. She recalls that one of her first works was already done with ice, material that offers a "fast way to have a shape becoming another thing or losing its shape." It questions the way territorialities imply orientation through the various dimensions of space and explore the consequences of their blurring. For her, the process of water changing states from liquid to ice and back reflects your feelings when, "for example, you have a text and you lose its meaning, or when you have a map and you have no element to interpret it [...]." The progressive disappearance of the sculpture alludes to the frailty of borders. The latter can thus be defined as based on power systems which are enforced by their existence in a self-produced dynamic which a random event could unbalance.

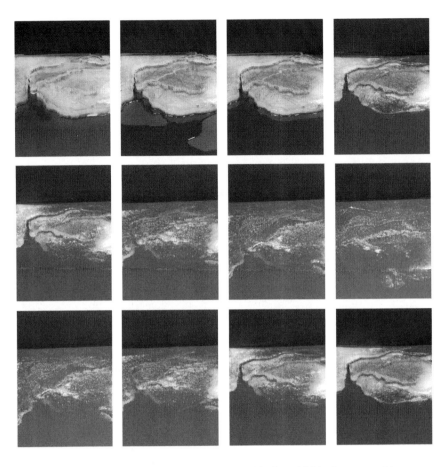

**Figure 10.3   Marina Camargo, *Tratado de Limites*, 2011, the carved ice
Mercosul borders are melting**

Source: Photos by Marina Camargo, with the permission of the author.

The juxtaposition of these elements is not easy to justify and their author tries
to escape this issue: "I am not very concerned to present a coherent position but
I want to approach the closer I can to my ideas." We cannot totally believe her in
those last words since indeed her living on the limits makes up for a very profound
sensible experience of the kinds of assemblage that define contemporary borders.
Camargo's *Tratado de Limites*, while intimately rooted into the Rio Grande do
Sul's landscape, still attains a very universal expression of what border work
consists of. The contrasts between the visibility and invisibility of the line(s), their
multiplicity and uniqueness, the subtle interrelations between crossing and control
are all beautifully exposed here.

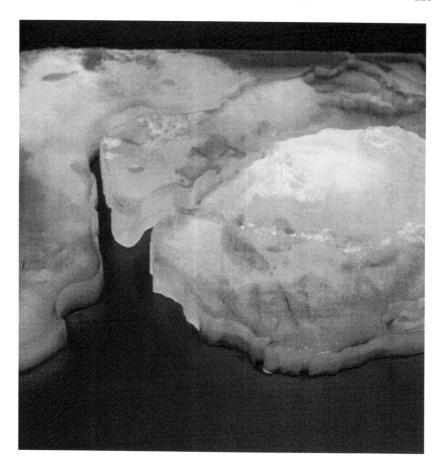

Figure 10.4 **Marina Camargo, *Tratado de Limites*, 2011, detail of the melting carved ice Mercosul borders**

Conclusion

Borders have not disappeared from the Brazilian art scene after the eighth Porto Alegre Biennial, although the 2013 edition switched to quite different spatial preoccupations, i.e. climate change. It is no coincidence that it also appeared in other contemporary events, such as the promotional show of contemporary Brazilian visual arts bearing the name of "From the Margin to the Edge. Brazilian Art and Design in the 21st century." The exhibition was displayed in London last summer (2012, 7/21 to 09/08), at Casa Brazil at Somerset House, and was commissioned by the Rio 2016 Olympic organizing committee. It accounts for the Brazilian strategy to become a global stakeholder through a visible world event. The 2014 World Cup in Brazil is another visible step toward achieving this globalization of Brazilian influence, but the contemporary art showcasing that goes with it is not to be underestimated. The title of the exhibition could be considered an oxymoron: it heralds the quest for a new central position by claiming peripherality, a contradiction that can only be overcome through the dialectical bordering process that artists who have experimented Brazilian borders can perform and share.

Expressing borders from the Porto Alegre art Biennial displayed some of the complexity of border places because both its curators and artists insist on the discomfort of their position:

> Partly traced over the center-periphery, the local-global opposition does not allow dichotomist readings either: its deconstruction demands a space for uncertain realms and precarious placements which consider opposition to be a contingent tension rather than a logical contradiction. Like a dispute that always makes reference to third positions. [...] We find that a biennial like this one, located in the Mercosur region, must necessarily render the problem more complex by introducing the third term of the regional. (Escobar in Escobar and Pérez-Barreiro 2007: 134)

Regions and borders alike, but maybe space in general, belong to this third space: neither here nor there, center or periphery, but constantly territorialized places where identity can be embedded and shared, and confronted. "Neither Center Nor Periphery" is a potentially very strong positioning, claimed in various discourses by Subcomandante Marcos who stressed that bringing back this kind of geography into politics was a powerful means of resistance. It is noteworthy that the Mercosul Biennial is visited by schools and groups, and the work shown there represented a genuine identity oriented pedagogical effort, aiming at embracing the complexity of the region with its future generations.

Let me give the last words to the eighth Mercosul Biennial chief-curator, José Roca. In a road map for the event that he traces in 20 points (his "[duo]decálogo"), the seventh point reads as such:

A biennial does not document (a)[8]

> If the work takes place in time, or outside the physical limits of the exhibition space, one has to let it live (and die) there. There is nothing more frustrating than an exhibition that documents performances, actions, temporary and outdoors works, which are presented to us as a reminder of what we were unable to

---

8   Documenta is the name of a contemporary art fair held in Cassel, Germany, every five years, and which is well considered within the contemporary visual arts scene.

experience. Unless it was conceived as a work, or has especially significant contextual value, documentation belongs in the archive, not in the exhibition. (Roca in Dias Ramos and Roca 2011: 23)

A detour through visual arts production for the purpose of a critical geopolitical understanding serves to highlight the performative dimension of borders. They exist because we experience them. Border work is not only top-down oriented, made up of security policies, control procedures and fear: the multiplicity of individual detours contribute as much to border design as the dominant stake-holders' agency. Their anonymity makes them very complex to describe. Artists should not be considered as transmitters of common representations. However, their capacity to engage with both conceptual and sensitive relations to space raises their testimony as precious component for those who wish to question the links between identity and place-building, and the complexity of territorialities.

References

Biennial Catalogues

Morgan, Robert C., Paternosto, César, Pérez-Barreiro, Gabriel. 2007. Francisco Matto: exposição monográfica, 6a BIENAL DO MERCOSUL realizada em Porto Alegre de 1º de setembro a 18 de novembro de 2007. This book is an integral part of the editorial project of the sixth Mercosul Biennial, Porto Alegre, September 1–November 18, 2007, Fundação Bienal de Artes Visuais do Mercosul, Porto Alegre, 2007, p. 221 (with parallel texts in Portuguese, Spanish and English). Available at: http://www.fundacaobienal.art.br/novo/arquivos/publicacao/pdf/Catalogo6B/Matto_catalogo_6B.pdf .

Escobar, Ticio, Pérez-Barreiro, Gabriel. 2007. Três fronteiras (6a Bienal do Mercosul), Porto Alegre: Fundação Bienal do Mercosul, 2007, p. 159 (with parallel texts in Portuguese, Spanish and English). Available at: http://fundacaobienal.art.br/novo/index.php?option=com_publicacao&Itemid=25&task=detalhe&id=34.

Dias Ramos, Alexandre (publication coordinator), Roca, José (general curator) and colaboração de Alexia Tala, Aracy Amaral, Cauê Alves, Fernanda Albuquerque, Pablo Helguera, Paola Santoscoy, 2011, 8a Bienal do Mercosul: ensaios de geopoética: catálogo, 2011, Porto Alegre: Fundação Bienal do Mercosul, 2011. Edição trilíngue (português, espanhol, inglês). Available at: http://pt.scribd.com/doc/98310058/Catalogo-8-Bienal-Mercosul.

General References

Amilhat Szary, A.-L. 2007. Are Borders More Easily Crossed Today? The Paradox of Contemporary Trans-border Mobilities in the Andes. *Geopolitics*, 12(1): 1–18.

Amilhat Szary, A.-L. 2010. Frontières et intégration régionale en Amérique Latine: sur la piste du chaînon manquant. *Regards croisés sur les intégrations régionales Europe/Amériques*, edited by C. Flaesch-Mougin and J. Lebullenger. Bruxelles, Bruylant, 307–41.

Amilhat Szary, A.-L. 2012a. Murs et barrières de sécurité: pourquoi démarquer les frontières dans un monde dématérialisé? *Dictionnaire des mondialisations*, 2nd Ed., edited by C. Ghorra Gobin. Paris: Colin, 447–51.

Amilhat Szary, A.-L. 2012b. Border Art and the Politics of Art Display. *Journal of Borderlands Studies*, 27(2), 213–28.

Amilhat Szary, A.-L. 2012c. The Geopolitical Meaning of a Contemporary Visual Arts Upsurge on the Canada/US Border. *International Studies*, LXVII(4), Special issue on "Canada after 9.11," 953–64.

Amilhat Szary, A.-L. 2013. L'artiste passe-muraille? La résistance au passage du mur entre Israël et la Cisjordanie (The Artist Walking Through Walls? Resistance and Barrier Crossings Between Israel and the West Bank). Available at: http://www.espacestemps.net/articles/lartiste-passe-muraille-la-resistance-au-passage-du-mur-entre-israel-et-la-cisjordanie-2/.

Aubertin, C. and Philippe, L. 1986. Frontières: mythes et pratiques (Brésil, Nicaragua, Malaysia). *Cahiers des Sciences Humaines*, 22(3–4), 254–480.

Beaverstock, J.V., Smith, R.G., and Taylor, P.J. 1999. A Roster of World Cities. *Cities*, 16(6), 445–58.

Brasil, Ministério da Integração Nacional. 2005. Proposta de Reestruturação do Programa de Desenvolvimento da Faixa de Fronteira. Brasília: Ministério da Integração Nacional: Grupo Retis/UFRJ. Available at: http://www.retis.igeo.ufrj.br/index.php/autor/rebeca-steiman/proposta-de-reestruturacao-do-programa-de-desenvolvimento-da-faixa-de-fronteira/#ixzz2g75ZozQF.

Brunet-Jailly, E. 2007. Theory and Policy: A Model of Border Security. *Borderlands: Comparing Border Security in North America and Europe*, edited by E. Brunet-Jailly. Ottawa: University of Ottawa Press.

Dorfman, A. 2008. Nacionalidade doble-chapa: novas identidades na fronteira Brasil-Uruguai. *A emergência da multiterritorialidade: a ressignificação da relação do humano com o espaço*, edited by A.Ln Heidrich, B. Pinós da Costa, C. Zferino Pires, and V. Ueda. Porto Alegre: Editora da UFRGS, 241–70.

Dorfman, A. 2012. Representações, normas e lugares: contos de contrabando da fronteira gaúcha. *Para Onde!?* (UFRGS), 6, 102–13.

Dorfman, A., Borba Colen, A., and Corseuil Duran, F.R. 2013. Barreiras comerciais, agências nacionais de saúde e o uso de agrotóxicos nos cítricos brasileiros. *Boletim Gaúcho de Geografia*, 40, 34–52. Available at: http://seer.ufrgs.br/bgg/article/view/3721.

Florida, R. 2002. *The Rise of the Creative Class and How It's Transforming Work, Leisure and Everyday Life*. New York: Basic Books.

Foucher, M. 1991. *Fronts et frontières. Un tour du monde géopolitique*. Paris: Fayard.

Foucher, M. 2012 [2007]. *L'obsession des frontières*. Paris: Perrin.

Godoi, R. and de Castro, S. 2010. Fronteiras abertas, Um retrato do abandono da Aduana Brasileira. Brasilia: SINDIRECEITA, Sindicato Nacional dos Analista-Tributarios da Receta Federal do Brasil.

Gruzinski, S. 1999. *La pensée métisse*. Paris: Fayard.

Gruzinski, S. 2004. *Les quatre parties du monde: histoire d'une mondialisation*. Paris: Ed. de la Martinière.

Hepple, L.W. 2004. South American Heartland: the Charcas, Latin American Geopolitics and Global Strategies. *Geographical Journal*, 170(4), 359–67.

Laplantine, F. and Nouss, A. 1997. *Le métissage*. Paris: Flammarion.

Laplantine, F. and Nouss, A. 2001. *Métissages, de Arcimboldo à Zombi*. Paris: Pauvert.

Ludmer, J. 2002 (1st Spanish edition 1988). *The Gaucho Genre: A Treatise on the Motherland Paperback*. Durham: Duke University Press.

Mignolo, W.D. 1991. *The Idea of Latin America*. New York: Wiley-Blackwell.

Oliveira, T.C.M. (Ed.). 2005. Território sem limites: estudos sobre fronteiras. Campo Grande, MS: Ed. UFMS. Fonte: Grupo Retis/UFRJ. Available at: http://www.retis.igeo.ufrj.br/index.php/ano/2005/territorio-sem-limites-estudos-sobre-fronteiras/#ixzz2g75TbiaI.

Novaes, A.R. 2009. Borders in the Press: South American Borders in Brazilian Journalistic Cartography. Paper to the Royal Geographic Society: Annual International Conference, Manchester.

Osorio Machado, L. 1998. Limites, fronteiras, redes. Proceedings of the III Coloquio Internacional de Estudios Fronteiriços. Sant'Ana do Livramiento, Brasil/Rivera, Uruguay, AGB (As. dos Geografos Brasileirios), Secao Porto Alegre, 41–9.

Perrier Bruslé, L. 2007. The Front and the Line: The Paradox of South American Frontiers Applied to the Bolivian Case. *Geopolitics*, 12(1), 57–77.

Roca, J., Tala, A., Amaral, A., Alves, C., Albuquerque, F., Helguera, P., Santoscoy, P. 2011. 8a Bienal do Mercosul: ensaios de geopoética: catálogo, Porto Alegre: Fundação Bienal do Mercosul. Available at: http://pt.scribd.com/doc/98310058/Catalogo-8-Bienal-Mercosul.

Rouquié, A. 1987. *Amérique latine: introduction à l'Extrême-Occident*. Paris: Seuil.

Rosière, S. 2013. La géopolitique au Brésil. Importation et réexportation d'une discipline controversée. *De Recife à Reims: récits géographiques, Mélanges offerts à Pernette Grandjean*, edited by M. Bazin, C. Fournet-Guérin, and S. Rosière. Reims: ÉPURE—Éditions et Presses universitaires de Reims, 155–85.

Thrift, N. 2008. *Non-Representational Theory: Space, Politics, Affect*. London: Routledge.

Chapter 11
"No Place Like Home":
Boundary Traffic through the Prison Gate

Jennifer Turner

Introduction

Similar to the identities constructed for non-nationals, the prisoner is often determined as the Other, and at a distance, metaphorical as well as physical, from the citizen majority. In this way, prisons and the diverse penal systems that they help make manifest, have entered the geographic imagination as prime exemplars of how seemingly invisible, peripheral sites are integral to the functioning of a purportedly mainstream society. However, despite their often peripheral physical locations, the inter-linkages between prisons and society are numerous and complex. Recent work within and beyond the discipline of geography acknowledges that where we might imagine a sharp boundary between the hidden inside and outside of prisons, there is in fact a myriad of materials that cleave and bind penal geographies, including forms of communication and inscription, networks of machines or technological devices, and buildings, which mark the prison walls as a site of transaction and exchange (Baer and Ravneberg 2008, Gilmore 2007, Loyd et al. 2009, Pallot 2005, Vergara 1995, Wacquant 2000, 2001, 2009).

Political economic analyses stress the profit-generating potential that prisons offer, with the prison system representing a "recession-proof economy," which is further entrenched by capitalism's need to enforce pauperism and criminality (Bonds 2006, Dyer 2000, Lemke 2001, Neumann 2000, Venn 2009). Indeed, the prison can be seen to provide a venue for entrepreneurial investment, "solving" the economic problems of depressed regions (Che 2005, Coulibaly and Burayidi 2010, Daniel 1991, Farrington and Parcells 1991, Forest County Conservation District and Planning Commission 1998, Gilmore 2007, Hooks et al. 2010, Millay 1991, Pickren 2011, Rollenhagen 1999), such that prisons now act as "geographical solutions to political economy crises" (Gilmore 2007: 26). Peck, for example, argues that in a neoliberal economy the prison system is not located on the "edges" of society—"as the designation 'boundary institution' might imply" (2003: 227)—insofar as it has become a site of privatization and commodification of services more traditionally associated with state welfare. There is a penalization of poverty here, Wacquant (2009) argues, wherein the (especially) urban disorders wrought by economic deregulation are met by what he calls a "garish theatre of civic morality" that targets deviant figures, such as the welfare mother and the teenage thug, as beyond both social and economic protection by the state. For Foucault (see Foucault 1986), prison is a heterotopia—a space that mirrors, yet subverts everyday life. According to Foucault, a heterotopia is the manner in which society and culture define the subject through his differentiation from general society, creating identity categories from the normal to the abnormal. On the one hand, Foucault considers individuals as free-willed members of a society, but at the same time they are subjects of a culture that examines, labels, and constructs them. This reconstitution of identity is one way of disciplining the subject and regulating behavior. Furthermore, Foucault argues that heterotopias are almost invisible and perceived as natural by members of a society. Yet, they are nevertheless measures for controlling and punishing the deviant or different—constituting a critical process in the formation of societal norms and expectations. Indeed, these binary geographies between inside and outside register as ideological obfuscations that hide the crucial role of prisons in current society. The prison is therefore an example of one of several boundaries marked out by the state.

When engaging with the distinct population of the prisoner, I acknowledge a plethora of novel power relationships that ensue as this quite particular boundary between outside and carceral space is crossed, and often blurred. This boundary is different from that crossed by other populations—in scale, legality, expectation, etc.—meaning that the prisoner allows us to move away from the typical populations encountered in border studies, in order to unravel the numerous scales, the differing boundaries and multiple power-space geometries that operate when different types of people move across and between variously defined territorial/legal borders. This consideration of the prisoner as an agent in border-crossing is situated alongside geographic research that provides a more nuanced understanding of other unique bordering practices. Examples include the bordered lives of young people (Aitken and Plows 2010); everyday experiences of home-making in relation to nature in suburban homes (Power 2009); embodied experiences of negotiating the boundary between the physical and the virtual world of video games (Ash 2010); and the theorization of geopolitical structural space in comic books (Dittmer 2007). In this chapter I similarly explore a unique border negotiation, namely prisoner employment as the cornerstone of successful rehabilitation, and examine the implications of this type of boundary interaction upon prisoner identities. I ground this discussion in examples drawn from research conducted on employment schemes for offenders who are either in prison and employed outside on day release, or employed within a company while on probation. By examining these cross-prison-border practices, I examine the implications of belonging to a group of conventional employees *and* those with criminal records; revealing a careful negotiation of "inside" and "outside" identities that complicate matters of belonging. In response to this, I explore the unwanted, or less than ideal, relationship with the prison as home, based on the lack of ability to re-integrate with the community that prisoners re-enter.

Throughout the analysis, this chapter attends to the hybridity of ex-offender constructions of "home." I illustrate the strong ties to prison and its problematic relationship with the high levels of recidivism that Britain is currently experiencing. A successful outcome for prison service authorities would be that, following completion of a prison term, ex-offenders would leave prison—embarking on a one-way journey. However, this chapter argues that, although ex-offenders may idealize a return to the communities where they lived prior to incarceration, the ability to re-integrate is often limited owing to the transformations individuals undergo following imprisonment. For example, this may happen consciously, such as through programs designed to shape psychological behavior; subconsciously, owing to friendships and allegiances developed within prison; or becomes manifested in legal markers like the possession of a criminal record. The body, in effect, becomes the carrier of the border (Amoore 2006: 347–8) acting as the finest scale of political space (Hyndman and de Alwis 2004: 549). This concept also displays similarities with offenders released from prison on license, or with electronic tagging, where the body literally carries a marker with them. As such, prisoners' everyday lives engage with both a physical and metaphorical boundary between prison on the inside and non-prison outside that performs a unique type of border crossing—that serves to create, re-imagine, blur, and even ignore this border. In this way, the prison wall becomes a porous boundary, with the prison gate facilitating two-way traffic across it. In considering this type of border work, this chapter responds directly to the current call for attention to the value of perspectives from cultural geography in attending to the nuanced space of the prison, and in drawing out their significance "to open up the political at a more 'personal' level" (Turner 2013: 35).

Examining the Prison/Non-prison Boundary

The purpose of the prison is to remove those deemed a threat to places that exist beyond liberal territory—which is where the literatures attending to borders and nations begin to dissipate into the field of interest of carceral geographies. Anderson (1996) argues that border control—the effort to restrict territorial access—has been a key concern of the state for a long time. Borders are traditionally viewed first and foremost as "strategic lines" which could be defended or breached by the military (Waltz 1999). According to Krasner (1985), it is the right of all states to determine who and what is granted legitimate territorial access. While the movement of trade and commerce is widely (and often necessarily) eroding the barriers between states, recent global events simultaneously shifted the emphasis away from military border concerns towards stronger policing of borders to maintain territorial control by each individual state—framing boundaries as social processes and practices that marginalize different geopolitical groups (Berg and van Houtum 2003, Newman 1999, Paasi 1996, 1999, Sundberg 2008: 876).

Physical barriers are created to prevent the movement of armies, uneconomical trade, as well as perceived "undesirables"—including migrants and the criminal (Andreas 2003). Border enforcement, imprisonment, *and* criminalization are fundamental to nation-state building as they are "also key technologies in the continuous processes that make up citizens and govern populations" (Pratt 2005: 1). However, Loyd et al. "challenge the idea that borders and prisons create safety, security, and order" (2012: 3) and posit them as a problematic method of categorizing people as, for example, "illegal" or "criminal." Post-9/11 governments suggest that "because they are fighting an unconventional enemy that has the capability to strike at any time and anywhere, governments need exceptional powers to prevent future attacks" (Jones 2009b: 880). As Andreas (2003) maintains, borders are not eroding or remaining unchanged, but are being re-crafted though new state regimes to exclude certain groups, while assuring territorial access for "desirable entries" (see also Sparke 2006). In this case, sovereign power does not only operate at the border. There are many efforts to observe, sort, and characterize people in their everyday lives, such as CCTV, biometric passports, immigration policing, visas, and so on, all of which aim to mark bodies as either acceptable or as a threat (Fyfe and Bannister 1996, Jones 2009a). We are also categorized, for example within UK populations, through the implementation of mechanisms such as Anti-Social Behavior Orders (ASBOs) or public house banning orders, both of which are designed to mark those who are deemed to be a threat to the rest of society.

Ridgley (2008) describes how, eight weeks after 9/11 took place, US Attorney General John Ashcroft authorized the questioning of around 5,000 Middle Eastern men, many on temporary visas in the United States. Although they were not suspected of any crime, they were selected because they matched "criteria of persons who might have knowledge of foreign-based terrorists" according to their gender, age, and national origin (Office of the Attorney General 2001: 1). The unlawful action of the government based on stereotypes was called into question by civil liberties and immigrant rights groups who were concerned about racial profiling, mass detentions, and the targeting of immigrant communities in the post-9/11 security climate. Furthermore, Guantánamo Bay, Cuba—a detention center for those suspected of terrorist activities, which suspends the rule of law—bears striking similarities to the concentration camps holding Jews, Romani Gypsies, homosexuals, disabled people, and many others, run by Nazis in World War Two. Guantánamo Bay acts as a cornerstone for a regime in which liberal movement and multiculturalism poses a threat to the political authority of the state (McClintock 2009).

As such, strategies aimed at spatial exclusion are proliferating across contemporary societies. In the US, over 1.3 million people were in state or federal prisons in 2000, up from 218,000 in 1974 (US Department of Justice, Bureau of Justice Statistics 2000: Table 6.27). This massive expansion of the prison system is a strategy that is being emulated by other countries with high numbers of prisoners, such as China and the UK. The processes of locking people up, and the segregation that it entails, renders incarceration a common and everyday practice (Gill 2009: 187). Yet, this is the obvious method of segregating space and creating boundaries within a national space. Other literature concerning the regulation of the built environment calls for focus upon architectural forms of socio-spatial exclusion, such as the methods used in urban spaces to channel social undesirables. These include the fences, gates, walls, surveillance, and armed security which protect luxury areas in cities such as São Paulo, London, or Los Angeles (Caldeira 2000, Crawford 2008, Davis 1998, Harcourt 2005, Lynch 2001). Furthermore, previously-lost methods of criminal justice such as banishment are being re-imagined as concepts in the contemporary world (Beckett and Herbert 2010), and other mechanisms such as sex offender registration (Tewksbury 2002) emerge as hybrid tools to control populations. In Seattle, for example, the adoption of "civility codes" and "trespass admonishments," which prevented panhandling, sitting on sidewalks and camping; and "drug-free zones" and "prostitution-free zones" imposed in Portland, Oregon compelled certain people to avoid certain places for a period of time. Prior to the implementation of these ordinances, authorities needed probable cause, such as a criminal offence, to remove individuals from public space. Now, the Seattle authorities authorize police and parks officials to exclude alleged rule violators without providing any evidence of wrong-doing—effectively "banishing" the homeless from public spaces (Beckett and Herbert 2010: 6). As Beckett and Herbert (2010) rightly identify, the rationale of banishment divides the world neatly into orderly and disorderly, identifying the latter spaces as those rife with crime and vice.

However, in spite of these efforts, concerns about the effective segregation of law-breakers from the rest of society derive from the presentation of the physical boundary of the prison as an exchange point. The prison walls also act as a border for other transactions, both legitimate and illegitimate. A prison sentence does not usually equate to complete isolation. Ties to the outside world are allowed, and are promoted by the flow of both people and objects such as gifts, letters, and photographs. However, there are also many other illegal interactions such as the presence of contraband mobile phones or drugs, recruitment of gang members, and even escape attempts, which illustrate that the border is neither as solid nor regulated as the directors of the penal system would hope. In response to this, in terms of the prison itself, location is often used to further deepen the boundary between carceral space and the rest of society.

Dartmoor prison located on barren moorland in Devon, England remains as one of these examples, as does Peterhead Prison, a lonely granite fortress in Scotland towering above a crag into the North Sea many miles from the nearest town (Sparks 2002). The architecture of the buildings themselves is often used to further enforce the prison/non-prison divide. Traditionally, prisons exhibited a threatening exterior, which was often decorated by gargoyles or figures pictured behind bars. This meant that as well as the bricks constituting a physical boundary, the public could still be reminded of the somber nature of the building and the detrimental aspects of committing crimes through the metaphorical juxtaposition with law-abiding society (Pratt 2002: 37).

Acknowledgment of the importance of spatial boundaries manifested itself in the architecture within prison as well. The "separate system" relied heavily upon penal architecture to enforce a regime of total cellular separation

(Henriques 1972, Ogborn 1995: 301). This system followed a well-defined rationale about particular ideas about "society" and the "individual." As well as concerns about poor ventilation, lack of hygiene and the spread of "gaol fever" for example, depravity (corrupt acts or practices) was considered to be something that could cause some kind of moral infection or contamination. By using architecture and separating practices in the daily regime to reduce contact, both physical and moral contamination of new prison arrivals by more "hardened" criminals could be eliminated (Fiddler 2010, Ogborn 1995: 301).

Aside from physically separating prisoners from law-abiding society, one of the main aims of imprisonment is to furnish individuals with skills that enable them to manage the transition from prison to society successfully upon release. In the remainder of this chapter, I explore offender employment programs as one mechanism that penal authorities implement in order to facilitate the successful re-integration of prisoners with outside society. As I will argue, there are many occasions whereby prisoners gain skills that enable them to effectively—that is, permanently—cross the prison border. However, what is also apparent, is a number of attachments to the prison "home" that allow us to posit the prison walls as a two-way boundary, across which individuals often participate in a cycle of leaving and return.

Working *With* and *In* the "Outside"

One of the methods by which offenders and those released from prison can reassert some of their liberties, perform obligations of good citizenship, and therefore successfully negotiate a permanent crossing to the world outside of prison is in employment. In order to ground this discussion, I draw upon research carried out with two different employment schemes specifically targeting prisoners at various stages in their sentences.[1] The first is the Oxford Citizens Advice Bureau (hereafter OxCAB), which employs prisoners who are permitted daily release from HMP Springhill to work as prisoner-advisers. Secondly, I discuss Blue Sky Development and Regeneration, which employs those released from prison on a six-month paid contract. In doing so, I critique the method in which schemes designed to allow prisoners and ex-offenders training in work-based environments merely serve to embody wider contemporary work ethics. These employments also reflect societal

---

1   A number of different companies targeting offender employment as a means of rehabilitation were approached via email or written correspondence. Those detailed in this chapter represent two case studies where I received invitation for further research. Original intentions were to recruit participants from those currently incarcerated *within* UK institutions to research prisoner attachment to "home." However, due to the legal and ethical issues surrounding access to this environment—chiefly the prisoners' inability to give non-coerced informed consent— this chapter uses empirical evidence obtained from two of the companies that responded to my call for participants.

expectations of modern citizenship (in general) and the mechanisms via which penal authorities negotiate a particular rendering of these as they seek to create, recreate, and reform "ideal" workers/citizens who are, or were, subject to a penal system of punishment and rehabilitation.

OxCAB developed the idea of training prisoners to become volunteer citizens' advisers in order to increase capacity and meet growing demand. The Citizens Advice Bureau (CAB) delivers advice services from over 3,500 community locations in England and Wales, run by 382 individual charities. The membership organization of the bureau is run by Citizens Advice, which is itself a registered charity. Following much debate, prisoners at HMP Springhill are now able to become Citizen Advisers. Springhill is one of the country's 12 category D open prisons,[2] to which male prisoners at the lowest level of risk are allocated. It does not take sex offenders or arsonists. Owing to the fact that people who visit the CAB are also often vulnerable members of society, the selection of prisoners is rigorous and they must comply with certain eligibility requirements. Alongside other volunteer advisers, their training and performance monitoring is on-going (interview, facilitator, January 12, 2012). Prisoners are available on a full-time basis and typically work four times as many hours as other volunteers. As such, most prisoner volunteers complete their training more quickly than traditional volunteers (within 6–8 weeks) and the OxCAB can now give advice to many more people than ever before.

Research was also conducted with Blue Sky Development and Regeneration, in the Wakefield branch located in West Yorkshire in England. Working as a viable business venture, Blue Sky tenders for commercial contracts from soft-landscaping (designing elements of a landscape such as fencing and planting) through to recycling and ground-working (digging foundations and other under-support for various types of structures). The company reinvests income into providing six-month contracts specifically for those with a criminal record who are involved with their local Probation Trust. Originally designed as a rehabilitative scheme through the Future Jobs Fund, Blue Sky has developed into a profitable company, with schemes such as recycling plants generating profits for the Local Authority of £120,000 per annum. However, more importantly alongside the work experience, employees are supported in CV building and given the opportunities to do training courses for such things as construction-machinery operation and building-site safety, each costing as much as £800 per person (interview, facilitator, August 11, 2011).

In the first 18 months of the project, Blue Sky Wakefield helped 19 ex-prisoners successfully finish their employment contracts. For many, this type of work was a new challenge, unlike any work they had carried out before. But, what is overwhelmingly apparent from those who I spoke to, is the ethos of care and

---

2 Category D prisons are known as "open" institutions, offering much greater levels of freedom including day release. Category A and B prisoners present the greatest security risk, whereas individuals who are categorized C prisoners are those who cannot be trusted in open conditions but are unlikely to try to escape.

future well-being that is promoted by the project. Facilitators highlighted the main focus to be getting employees back onto the first rung of the ladder to a stable routine of work and earning a steady income—a package attractive enough to prevent them from reoffending.

In both the cases of Blue Sky and OxCAB, focus groups were carried out in the workplace with participants who had volunteered and been selected by the company themselves.[3] Conversations were taped, with permission from the participant, and later transcribed and coded using the key themes of my wider doctoral research focusing upon engagements across the prison/non-prison boundary—namely penal spaces, penal identities, rehabilitation, and constructions of "home" (either on the "inside" or the "outside"). Participants' names are anonymized, although the case studies of Blue Sky and OxCAB are not, as identification was requested owing to the companies setting a precedent for other offender/ex-offender employment schemes, and interviewees agreed to be anonymized in conjunction.

Crossing the Border to the "Outside"

As findings by Graffam et al. (2005) suggest, gaining paid employment upon release lowers the rate of reconviction. Therefore, getting a job can be highly significant in helping ex-offenders reintegrate into society outside of prison. My interviewees recognize that contributing to tax, National Insurance and paying their way, rather than society paying for them, all play a part in their process of normalization. Barke (2001) notes the emotional importance of dwelling in a private, domestic property—allowing individuals the freedom from the critical gaze of society. In a similar way, employment generates an ability to achieve both reintegration into the mechanisms of capitalism, *and* the respect of family members through the wages it provides. At OxCAB, prisoner-advisers relish the feeling of "fitting in" with other, non-prisoner colleagues, and express a great enjoyment at having a conventional "9 to 5" working day outside of the prison environment. Indeed, the importance of achieving regular work in the future in order to legitimately provide for themselves and their families is a fundamental concern of both groups of participants. As the employees at Blue Sky explain:

> Ben: It makes it look better, doesn't it, when the family's saying 'oh what are you doing now?'
>
> Steve: It makes you feel better doesn't it? ... Yeah, cause you can say 'yeah I'm working now.' 'Aw right good, where are you working?' Instead of saying, 'oh yeah, [nothing], just the usual on the dole, sat at home.'

---

3   All of my participants were male. HMP Springhill is a male prison. Furthermore, although Blue Sky does also employ female ex-offenders, none were available to participate in the interviews.

Chris: Yeah it gives you that image that you're going out and working I think …

Rich: You feel better within yourself as well … looked up to by my missus and my baby … And you can afford things, you can afford to actually go out and do things … buy my baby things, and buy him nice clothes rather than having nothing. (Focus Group Blue Sky, August 11, 2011)

Although the OxCAB–Springhill partnership was originally developed to aid the Bureau rather than act as a rehabilitation project, prisoners find many benefits in their preparation for release. For the prisoners interviewed, certainly, work at the CAB is very different to anything they had ever done before: their role gives them the possibility of a sense of normality and social inclusion, particularly in relation to the expectations of family and friends. The impact that a positive family relationship can have upon the reduction of recidivism is widely attended to by both scholars across a variety of disciplines (Comfort 2002, 2008, McGarrell and Hipple 2007, Mills and Codd 2008) and official reports (Her Majesty's Inspectorates of Prisons and Probation (HMIPP) 2001, Home Office 2004, 2006). Once solid links exist they can be a major contributor to severing all ties with prison. Mills and Codd (2008) find that families generate "resourceful social capital," which can aid in forging positive links with liberal society—particularly useful when finding gainful employment (Farrall 2004). After a string of cautions, Chris finally received a prison term, which he claims, has scared him. However, he worries that the relationships he has built up since his release would simply be destroyed if he were to be incarcerated again:

I think they just gave me [my sentence] to make me realize … an eye-opener, just to scare me and I think it has really … I was scared shitless to be honest … but no, I don't want to go back. My girlfriend's pregnant … part of me going to jail was for pinching from [a supermarket] and my girlfriend's mum got me the job … she hasn't spoken to me for like 12 months and now we go to her house … so now, I'm building up a better relationship with her part of the family … if I went back to jail it would be for them to think 'what are you doing with him?', you know what I mean? (Focus Group Blue Sky, August 11, 2011)

Conversely, 25-year-old Ben spent nine years in and out of prison. Similarly for him, the return to prison is undesirable as he now believes his life is much better outside of it and is motivated by the desire to perform his obligations to his family. He (like two of his other colleagues) believes that the birth of his first child changed the pattern:

Before it was just girlfriends I had to think about … but now I've got a kid, this time I went to prison all I could think about was I'd let her down, not just myself, I've let her down as well cause I'm not there to see her, you know what I mean, and look after her … I wouldn't like to go to prison now … it's alright when you've got something to do during the day, which isn't much to do, but it's something … it's when its night-time and you're just sat there with [nothing] to do and all you're doing is thinking and the only thing that kept coming into my head was my daughter … thinking I wanted to go home because I wanted to see her, and I wouldn't want to go again just to have that feeling again, that gut feeling again because it was horrible. (Focus Group Blue Sky, August 11, 2011)

Aside from the reward of praise from family and friends (and the stronger familial ties associated with this), these types of employment stimulate something else amongst prisoners: a sense of personal aspiration. A comparison is sometimes made with this "intellectually-challenging" employment and "mundane" or "mediocre" jobs (repetitive and low-skilled tasks such as laundry work) that are typically deemed appropriate for the prisoner:

Liam: I've never been in this line of work … building sites, warehouses, things like that, but … I've now discovered … that it is something that I enjoy doing, compared to what I was doing before, which was something that I didn't like doing … so, yes … I've discovered something new about myself … I didn't think I was into this sort of thing but now, obviously, I've changed completely.

Oliver: What I like about this here is that it's not mediocre work … Some of the community work jobs that people do are a bit … mundane if you like, they have to steam and press clothes, that sort of thing. (Focus Group OxCAB, January 12, 2012)

The CAB also aims to equip people with the knowledge required to deal with any problems they face, as well as endeavoring to shape the way policies influence daily lives. According to Citizens Advice, their service provides "free, independent, confidential and impartial advice to everyone on their rights and responsibilities" (2012, no page). It is interesting that in this case, then, that the OxCAB is staffed by prisoners, who society posits as *not good* citizens (Thompson 2000). And, interviewees spoke about the development of their personal knowledge database as a result of their work experience; particularly their ability to find sources of support and information for themselves following future release. Facilitators explained scenarios where prisoner-advisers are asked to help with form filling and make suggestions about appropriate organizations for their fellow inmates. In this respect, it can be argued that this turns the CAB itself into a training ground, both in terms of the users and the prisoner-advisers themselves.

Both of these examples illustrate the different methods by which prisoners may negotiate a successful, permanent border crossing from carceral spaces to liberal society outside. However, not all prisoners effectively enact these methods and there are a variety of reasons why certain offenders will return to prison. In the next section of this chapter, I address a number of barriers to re-integration (that may prompt a return journey through the prison gate), namely, prisoners' attachment to the prison "home."

The Prison "Home:" A Barrier to Re-integration

According to figures from the Ministry of Justice, in the quarter ending September 2011, over 32,000 first-time receptions were made to the prison system in England and Wales (see Table 11.1). During this same time, there were also nearly 22,000 offenders discharged from determinate sentences. Of those, 2,500 had served more than four years in prison. This means, that in addition to the prison/society relationship being highly fluid with numerous networks, there is a large number of people re-entering society having spent a significant time incarcerated within the specific prison lifestyle.

Table 11.1   Prison reception and discharge figures for England and Wales, quarter ending September 2011

Number of first-time receptions	32,212
Number of discharges	21,677
Discharges after sentences > 4 years	2,560

Source: Ministry of Justice (MoJ) 2011.

There is a very clear relationship between recidivism and high levels of ex-offender unemployment. Ex-offenders face many difficulties when seeking employment including poor reading, writing and numeracy skills; behavioral and health problems; debt and homelessness; as well as discrimination by employers (Webster et al. 2001). According to Opperman, 50 percent of prisoners cannot read and write or do basic mathematics. Hence, half of the prison population is lacking in the skills required for 96 percent of jobs (2012: 24). Unemployment among offenders is very high. A 1997 UK study found that employment of offenders under probation was only 21 percent compared to around 60 percent of the general population (Mair and May 1997). Fletcher et al. (1998) reported that ex-prisoners are said to comprise between two and three percent of the average monthly inflow to the unemployment pool. Aside from the personal issues that lack of a job may create, research highlights the impact of this upon reoffending. Although there is little direct evidence of a causal relationship between unemployment and crime, studies at the individual level reveal that periods of unemployment are related to periods of offending (Farrington 1996) and reconviction (May 1999). Currently, recidivism in the UK is of serious concern,[4] with more than a third of those released

---

4   Although it might be useful, raw reoffending rates between countries should not be directly compared as there are a range of underlying differences in the justice systems and the methods of calculation (See Ministry of Justice (MoJ) 2010. *Compendium of Reoffending Statistics and Analysis: Ministry of Justice Statistics Bulletin*. London: Home Office).

from prison committing another offence within the first year[5] (Ministry of Justice (MoJ) 2011). As such, the prison wall can be posited as a kind of border, with both "migrants" and "returnees" crossing in both directions for different lengths of time.

Furthermore, with many jobs in the UK requiring a criminal records check, the offence is likely to have significant logistical and legal effects upon the life of the ex-offender—acting as a permanent reminder of their incarceration. For many prisoner volunteers, the reference from OxCAB is a major contributing factor to their decision to participate, as it goes a long way to prove their credentials to potential employers, particularly in the face of the decreased opportunities owing to their criminal record:

> Liam: … the fact that it's good for future references … you're going to do well to get a reference when you leave prison, because obviously you've got the thing of having a criminal record, coming out of prison to get a job with employers not wanting to take you on, so obviously if you've done this you've got a head start, haven't you? (Focus Group OxCAB, January 12, 2012)

The potential negative consequences of being exposed as a former prisoner were raised as a concern by many participants. For example, it is clear that there remains a necessity for anonymity for staff who are prisoners to be retained with regards to members of the public using the Bureau. Although it is advertised that the branch acts in partnership with Springhill, the prisoner-advisers have never been individually identified as such to any clients. When asked to comment on anonymity and disclosure, the comments were varied:

> Oliver: I think it's quite important, but not to the people who volunteer … they might not like prisoners giving them advice.
>
> Liam: I don't mind them knowing. They might look at it as … you're giving back and you're changing your life around. (Focus Group OxCAB, January 12, 2012)

Although CAB staff recalled instances where the media were critical of the use of prisoners as advisers (with one referring to the "Con Advice Bureau"), facilitators maintain that the prisoner status is of no more relevance to the job role than someone's age or sexuality. This is largely based on the quality of work that CAB receives from its prisoner volunteers, the positive feedback, letters of thanks, and even monetary contributions, which members of the public offer in return for the service. Certainly, the invisible nature of the prisoner-adviser renders them able to

---

5   The report found that 39.3 percent of adults were re-convicted. This is defined as offences committed in a one-year follow-up period and convicted within the follow up period or a further six-month waiting period.

exert their participation in ways that may not be afforded by their exposure in this environment. However, it could be argued that the elision of volunteers' identities as prisoners serves precisely to erase the "prisoner" as "citizen" while ensuring that he/she must simultaneously give advice to others about how to successfully participate in civic regimes.

Conversely, my interviewees describe Blue Sky's workplace, where disclosure is one of the eligibility requirements, as one where the common ground is welcoming. No-one is forced to lie to anyone, as both employees and facilitators are aware of individuals having spent time in prison. Ben and Rich also commented upon the negative treatment they encountered when visiting the Job Centre to claim their Job Seeker's Allowance, and the difficulty in finding a job they have experienced.

> Rich: They treat you like you're trashy on that Job Seekers' [Allowance] ...
>
> Ben: ... you go in and they just talk down to you. They know you're on the dole, they know you're coming to sign on, you're getting your money for [nothing] ... They like really belittle you and talk down to you ...
>
> Rich: And they always say to you 'Why haven't you found a job, there's all these jobs out there?' ... but it's like you apply for hundreds and hundreds of jobs and they don't understand that not every job you are going to get. (Focus Group Blue Sky, August 11, 2011)

I note here that social processes of inclusion and exclusion critically depend on the categorization of people as belonging and non-belonging (Ralph and Staeheli 2011: 523). However, this categorization is less about the subjective feelings of the individual and more about powerful actors such as societal elites, and political authority figures, for example, saying who belongs and who does not (Castles and Davidson 2000, Crowley 1999, Ilcan 2002). Membership must be validated (Young 2011). Drawing upon Probyn (1996), ex-offenders clearly exist between two interrelated states that together define belonging: that of "*being*," and that of "*longing*." There is a definite antagonism between the actual and idealized meaning of home (Ralph 2009). It is unsettling for those released from prison to discover that, despite every effort to sever all ties to carceral spaces, they may no longer belong in the place they always called their *home*. The reality of return to life outside prison may be far from the one that was dreamt about. For some, this may create a disenchantment, but more importantly, forces others "to revise their self-identities and articulate a liminal status as both insiders and outsiders" (Ralph and Staeheli 2011: 523).

In recent years, geographies of home have come to be theorized as both material and symbolic, located on the threshold between past memories, the everyday present and future dreams and fears (Blunt and Varley 2004). In similar vein, Baer (2005) illustrates the manner in which prisoners "decorate" their cells with items considered mundane in the "outside," in order to provide a material link to the non-prison world. Equally, for migrants, the desire to pin down identities to a fixed home provides a stable sense of self in a world characterized by flux (Conway 2005). This flux is intrinsic to my on-going research surrounding the relationship, and more specifically the "boundary traffic," between prison and society (Turner 2013).

Scholars problematize notions of home as a fixed entity or physical dwelling place (Brettell 2006, Datta 2010). Instead, home is linked to local networks and communities, or even national identities through ideologies and practices with both humans and non-humans (Jacobs and Smith 2008, McDowell 1997, Miller 2001). Home is also a threshold-crossing concept, traversing the boundaries across time/ space. It is therefore messy, mobile, blurred, and confused (Ahmed et al. 2003, Nowicka 2007). For Ralph and Staeheli, "the challenge ... is to conceptualize the simultaneity of home as sedentarist and as mobile" (2011: 518). Therefore, the concept of home can be both dynamic *and* moored—a location, or a set of relationships that shape identities and feelings of belonging. Mobility *and* stasis, displacement *and* placement, as well as roots *and* routes go into the making of home (Clifford 1997, Gustafson 2001). This ambiguity about "home" is well researched, positing the possibility of multiple homes (Constable 1999, Ní Laoire 2007, 2008a, 2008b).

Recent work considers the generation of "hybrid" identities (Brubaker and Cooper 2000, Walter 2004, 2006, Yau 2007). Home, therefore, incorporates both a lived and longed for state (Ralph and Staeheli 2011: 522). Fluid, fragmented, or partial identities do not exclude the desire for an integrated, whole, and stable identification with home (Varley 2008, Young 1997). This is of particular interest when we consider penal spaces, and the generation of a hybrid attachment to both prison and the outside community they are released into. Participants also acknowledged the way that ex-offenders generate attachment to a prison "homeland," which results in further inability to forge positive links with the communities they are released into—making a return journey across the prison/ non-prison border all the more likely.

Scholars such as Hayner and Ash (1939) illustrate the informal rules created by inmate hierarchies, or gang allegiances, which exist alongside those of the administration. Other attachments may include adhering to prison jargon—which often becomes a subconscious activity (Fox 1999). They might become part of the system of supply and demand that is prominent in prison life, where everyday objects such as the foil in sweet wrappers become valuable trading commodities due to their alternative use as aids in drug-taking (Valentine and Longstaff 1998). This "inside" world soon becomes a domestic regime, a way of life, and in many cases a "home"—something that became clear to me even through the conversational language used when talking about the prison. On one occasion, an unintentional use of the word "home" sparked discussion:

JT: And so when you get home, oh sorry, I've said it again ...

Ian: [laughs]

Oliver: It's alright it is home.

Liam: I find myself saying that all the time ... when I'm on home leave, for instance D-cat prisoners can go home, I say to the missus or whatever, 'I've got to go home [back to prison] tomorrow,' I'm at home but I still say it. (Focus Group OxCAB, January 12, 2012)

However, for some, the ambivalence towards the prison environment is clear. Ben described to me how he settles fairly quickly into the routine of prison life, always easily achieving the most sought-after jobs, and learning to do what was necessary to "make it look good." The former-prisoners are quick to recognize the leniencies of the prison environment, with one describing it as "a boarding school where you just don't get to go home at the weekend." When I asked Ben if he wanted to go back he said not seeing his daughter and losing his job would be the only downside to it. Harman et al. (2007), for example, use evidence sourced from wives of incarcerated prisoners who are affronted and dismayed at the degree of free time and relaxation that their male partners enjoy when in prison, at precisely the time when they are having to manage both the family finances and the children themselves. There is also a clear recognition of the fact that some of the people who experienced prison found it to be less harsh than their original pre-conceptions. May and Wood (2005) demonstrate that many American prisoners would prefer to go to prison than do community service, house arrest, or "boot camp" when offered the choice. Furthermore, there are others who purposefully make prison a return destination if they are not succeeding in the outside world upon release. Prisoners can receive basic needs, such as shelter and food, but may also be offered opportunities not enjoyed by some people on the "outside," such as enhanced access to education (Cohen 2012) or a social network that they might lack elsewhere (Howerton et al. 2009). As such, a return journey to a homeland widely defined as problematic and undesirable may still exhibit appeal for this group of people.

For ex-prisoners, it seems they are torn. As discussed earlier in the chapter, prisoners may have families on the outside, often aiding their reintegration into liberal society. However, the friendships or "families" that are often metaphorically created in prison can also be strong—particularly for those with dysfunctional upbringings and other difficulties with their biological kin. This sense of ambiguity comes when prisoners exhibit a sense of allegiance with the other inmates. This sense of allegiance is something I explored with my interviewees insofar as it makes Blue Sky something of a nurturing environment; its employees can remain within the comfort blanket of like-minded people for the daunting and often-difficult first six months after release. Rich comments:

I don't know, there just seems to be something between people, because they've done the same kind of thing ... it's not like we start a new job and everyone's law abiding citizens and none of them have seen police unless they've phoned 'em, we're all the same so when we come to this job … you know that he's been in prison and he has, so you feel comfortable … (Focus Group Blue Sky, August 11, 2011)

Other recent work within a carceral setting notes particularly how prison may constitute a positive place of friendship (Caine 2006), or generate a hybrid form of prisoner citizenship (Turner 2012). Bronson (2008) observed the intense friendships that are forged within the prison environment, with commonalities provided by jobs on the outside, religions, birthplace or hobbies. These friendships also become intensified by the close contact of the penal environment, facilitating relationships as strong as familial ties:

Jake: Three or four guys in here I consider *almost like blood brothers*. Like they're real relatives. I know I could tell them anything, show them any side of me, whatever. (Bronson 2008: 79, emphasis added)

As such, these relationships further complicate the engagements by prisoners with the prison/non-prison boundary—creating a depth to the border that exists well beyond the prison gate—both blurring its solid definition and reproducing it as a two-way interface between the opposing sides.

Conclusion

In this chapter I consider the prison as one, among many, manifestations of a border. This is a boundary different to that crossed by other populations—in scale, legality, expectation, etc.—meaning that the prisoner allows us to move away from the typical populations encountered in border studies to consider the novel power relationships that ensue as this quite particular boundary between "outside" and "carceral" space is crossed (and often blurred). By illustrating the complexities of this situation in relation to prisoner employment programs I present one way in which we can re-conceptualize traditional notions of the border in order to unravel the numerous scales, the differing boundaries, and multiple power-space geometries that operate when different types of people move across and between variously defined territorial/legal borders.

In particular, this chapter considers that, while prison authorities would aim to produce ex-offenders that successfully negotiate a permanent border crossing from prison to society, high levels of recidivism in the UK are indicative of the number of individuals who are more likely to participate in a frequent cycle of leaving and returning. Prisoners and ex-offenders may generate a hybrid attachment to both prison and the outside community into which they are released. However, those like OxCAB and Blue Sky recognize the support of peers that ex-offenders may subconsciously require during a period in their lives where a sense of "home" might be ambiguous. Trapped between the place that they want to belong to and the one that binds them, the time spent in prison may indeed render them ever more absent from the societies they are released into, with their "prison home" remaining ever present in their everyday lives. The sentiment is worrying, as one interviewee commented, "Prison has totally changed me … but, deep down, you can never really leave" (Focus Group Blue Sky, August 11, 2011).

In view of this, however counter-intuitive they seem, prisoners and ex-prisoners may hold positive attitudes to prison, and this should be recognized by key agents in the penal system in order to produce a "person-centered approach to supporting resettlement" (Howerton et al. 2009: 458). In this way, perhaps paying attention to the hybridity of both "home" and prisoner-migrant may facilitate the reintegration from "inside" to "outside" more effectively, that is, a one-way journey through the prison gate to liberal society.

References

Ahmed, S., Castaneda, C., Fortier, A.M., and Sheller, M. (eds). 2003. *Uprootings/Regroundings: Questions of Home and Migration*. Oxford: Berg.

Aitken, S. and Plows, V. 2010. Overturning Assumptions about Young People, Border Spaces and Revolutions. *Children's Geographies*, 8(4), 327–33.

Amoore, L. 2006. Biometric Borders: Governing Mobilities in the War on Terror. *Political Geography*, 25(3), 336–51.

Anderson, M. 1996. *Frontiers: Territory and State Formation in the Modern World*. Cambridge, MA: Polity Press.

Andreas, P. 2003. Redrawing the Line: Borders and Security in the Twenty-first Century. *International Security*, 28(2), 78–111.

Ash, J. 2010. Teleplastic Technologies: Charting Practices of Orientation and Navigation in Videogaming. *Transactions of the Institute of British Geographers*, 35(3), 414–30.

Baer, L.D. 2005. Visual Imprints on the Prison Landscape: a Study on the Decorations in Prison Cells. *Tijdschrift voor Economische en Sociale Geografie*, 96(2), 209–17.

Baer, L.D. and Ravneberg, B. 2008. The Outside and Inside in Norwegian and English Prisons. *Geografiska Annaler Series B-Human Geography*, 90B(2), 205–16.

Barke, M. 2001. Housing, Space and Society. *Introducing Social Geographies*, edited by R. Pain, M. Barke, D. Fuller, J. Gough, R. MacFarlane, and G. Mowl. London: Arnold.

Beckett, K. and Herbert, S. 2010. Penal Boundaries: Banishment and the Expansion of Punishment. *Law and Social Inquiry—Journal of the American Bar Foundation*, 35(1), 1–38.

Berg, E. and van Houtum, H. 2003. *Routing Borders between Territories, Discourses, and Practices*. Burlington, VT: Ashgate.

Blunt, A. and Varley, A. 2004. Geographies of Home. *Cultural Geographies*, 11(1), 3–6.

Bonds, A. 2006. Profit from Punishment? The Politics of Prisons, Poverty and Neoliberal Restructuring in the Rural American Northwest. *Antipode*, 38(1), 174–7.

Brettell, C.B. 2006. Introduction: Global Spaces/Local Places: Transnationalism, Diaspora, and the Meaning of Home. *Identities—Global Studies in Culture and Power*, 13(3), 327–34.

Bronson, E.F. 2008. "He ain't my brother … he's my friend." Friendship in Medium Security Prison. *Critical Issues in Justice and Politics*, 1(1), 63–74.

Brubaker, R. and Cooper, F. 2000. Beyond "Identity." *Theory and Society*, 29(1), 1–47.

Caine, B. 2006. Prisons as Spaces of Friendships in Apartheid South Africa. *History Australia*, 3(2), 42.41–42.13.

Caldeira, T. 2000. *City of Walls: Crime, Segregation, and Citizenship in São Paulo*. Berkeley, CA: University of California Press.

Castles, S. and Davidson, A. 2000. *Citizenship and Migration: Globalization and the Politics of Belonging*. London: Macmillan.

Che, D. 2005. Constructing a Prison in the Forest: Conflicts over Nature, Paradise, and Identity. *Annals of the Association of American Geographers*, 95(4), 809–31.

Citizens Advice Bureau. 2012. *About Us. Citizens Advice Bureau*. [Online]. Available at: http://www.citizensadvice.org.Uk/index/aboutus.htm [accessed: June 1, 2012].

Clifford, J. 1997. *Routes: Travel and Translation in the Late Twentieth Century*. London; Cambridge, MA: Harvard University Press.

Cohen, S. 2012. Beyond Three Hots and a Cot: The Making of Places in Placeless Prisons. Paper to the Annual Meeting of the Association of American Geographers, New York, NY, February 26.

Comfort, M. 2002. "Papa's house:" The Prison as Domestic and Social Satellite. *Ethnography*, 3(4), 467–99.

Comfort, M. 2008. *Doing Time Together: Love and Family in the Shadow of the Prison*. Chicago, IL: University of Chicago Press.

Constable, N. 1999. At Home but Not At Home: Filipina Narratives of Ambivalent Returns. *Cultural Anthropology*, 14(2), 203–28.

Conway, D. 2005. Transnationalism and Return: "Home" as an Enduring Fixture and "Anchor." *The Experience Of Return Migration: Caribbean Perspectives*, edited by R. Potter, D. Conway, and J. Phillips. London: Ashgate.

Coulibaly, M. and Burayidi, M. 2010. Assessment and Removal of Prison's Effect on the Profiles of Rural Host Communities. Paper to the Annual Meeting of the Association of American Geographers, Washington, DC, April 17.

Crawford, A. 2008. From the Shopping Mall to the Street Corner: Dynamics of Exclusion in the Governance of Public Space. Paper to the Worldwide Universities Network Colloquium, International and Comparative Criminal Justice and Urban Governance, University of Leeds, UK, June 26–8.

Crowley, J. 1999. The Politics Of Belonging: Some Theoretical Considerations. *The Politics of Belonging: Migrants and Minorities in Contemporary Europe*, edited by A. Geddes and A. Favell. Aldershot: Ashgate.

Daniel, W.R. 1991. Prisons and Crime Rates in Rural Areas: the Case of Lassen County. *Humboldt Journal of Social Relations*, 17 (1&2), 129–70.

Datta, A. 2010. The Translocal City: Home and Belonging among East-European Migrants in London. *Translocal Geographies: Spaces, Places, Connections*, edited by K. Brickell and A. Datta. London: Ashgate.

Davis, M. 1998. *City of Quartz: Excavating the Future in Los Angeles.* London: Pimlico.

Dittmer, J. 2007. The Tyranny of the Serial: Popular Geopolitics, the Nation, and Comic Book Discourse. *Antipode*, 39(2), 247–68.

Dyer, J. 2000. *The Perpetual Prison Machine: How America Profits From Crime.* Boulder, CO: Westfield Press.

Farrall, S. 2004. Social Capital and Offender Reintegration: Making Probation Desistance Focused. *After Crime and Punishment: Pathways to Offender Reintegration*, edited by S. Maruna and R. Immarigeon. Cullompton: Willan Publishing.

Farrington, D.P. 1996. Criminological Psychology: Individual And Family Factors in the Explanation and Prevention of Offending. *Working With Offenders: Psychological Practice in Offender Rehabilitation*, edited by C. Hollin. Chichester: Wiley.

Farrington, K. and Parcells, R.P. 1991. Correctional Facilities and Community Crime Rates: Alternative Hypotheses and Competing Explanations. *Humboldt Journal of Social Relations*, 17(1&2), 17–127.

Fiddler, M. 2010. Four Walls and What Lies Within: the Meaning of Space and Place in Prisons. *Prison Service Journal*, 187, 3–8.

Fletcher, D.R., Woodhill, D., and Herrington, A. 1998. *Building Bridges into Employment and Training for Ex-Offenders.* York: Joseph Rowntree Foundation.

Forest County Conservation District and Planning Commission. 1998. *Comprehensive Plan.* Tionesta, PA: Forest County Conservation District and Planning Commission.

Foucault, M. (translated by Miskowiec, J.) 1986. Of Other Spaces. *Diacritics—A Review of Contemporary Criticism*, 16(1) 22–7.

Fox, K.J. 1999. Changing Violent Minds: Discursive Correction and Resistance in the Cognitive Treatment of Violent Offenders in Prison. *Social Problems*, 46(1), 88–103.

Fyfe, N.R. and Bannister, J. 1996. City Watching: Closed Circuit Television Surveillance in Public Spaces. *Area*, 28(1), 37–46.

Gill, N. 2009. Governmental Mobility: the Power Effects of the Movement of Detained Asylum Seekers around Britain's Detention Estate. *Political Geography*, 28(3), 186–96.

Gilmore, R.W. 2007. *Golden Gulag: Prisons, Surpluses, Crisis, and Opposition in Globalizing California.* London: University of California Press.

Graffam, J., Shinkfield, A.J., Mihailides, S., and Lavelle, B. 2005. *Creating a Pathway to Reintegration: The Correctional Services Employment Pilot Program (CSEPP): Final Report to Department of Justice.* Melbourne: Deakin University.

Gustafson, P. 2001. Roots and Routes. *Environment and Behavior*, 33(5), 667–86.

Harcourt, B.E. 2005. Policing L.A.'s Skid Row: Crime and Real Estate Development [An Experiment in Real Time]. *University of Chicago Legal Forum*, 325–404. Available at: http://papers.ssrn.com/sol3/papers.cfm?abstract_id=739130 [accessed: October 1, 2012].

Harman, J.J., Smith, V.E., and Egan, L.C. 2007. The Impact of Incarceration on Intimate Relationships. *Criminal Justice and Behavior*, 34(6), 794–815.

Hayner, N.S. and Ash, E. 1939. The Prisoner Community as a Social Group. *American Sociological Review*, 4(3), 362–9.

Henriques, U.R.Q. 1972. The Rise and Decline of the Separate System of Prison Discipline. *Past and Present*, 54, 61–93.

Her Majesty's Inspectorates of Prisons and Probation (HMIPP). 2001. *Through the Prison Gate: A Joint Thematic Review by HM Inspectorates of Prisons and Probation.* London: Home Office.

Home Office. 2004. *Reducing Re-offending: National Action Plan.* London: Home Office Communication Directorate.

Home Office. 2006. *A Five Year Strategy for Protecting the Public and Reducing Re-offending.* London: Home Office.

Hooks, G., Mosher, C., Genter, S., Rotolo, T., and Lobao, L. 2010. Revisiting the Impact of Prison Building on Job Growth: Education, Incarceration, and County-level Employment, 1976–2004. *Social Science Quarterly*, 91(1), 228–44.

Howerton, A., Burnett, R., Byng, R., and Campbell, J. 2009. The Consolations of Going Back to Prison: What "Revolving Door" Prisoners Think of their Prospects. *Journal of Offender Rehabilitation*, 48(5), 439–61.

Hyndman, J. and de Alwis, M. 2004. Bodies, Shrines, and Roads: Violence, (Im)Mobility and Displacement in Sri Lanka. *Gender, Place and Culture*, 11(4), 535–57.

Ilcan, S. 2002. *Longing in Belonging: The Cultural Politics of Settlement.* New York, NY: Praeger.

Jacobs, J.M. and Smith, S.J. 2008. Living Room: Rematerialising Home. *Environment and Planning A*, 40(3), 515–19.

Jones, R. 2009a. Agents of Exception: Border Security and the Marginalization of Muslims in India. *Environment and Planning D-Society and Space*, 27(5), 879–97.

Jones, R. 2009b. Categories, Borders and Boundaries. *Progress in Human Geography*, 33(2), 174–89.

Krasner, S.D. 1985. *Structural Conflict: The Third World against Global Liberalism.* Berkeley, CA; London: University of California Press.

Lemke, T. 2001. "The birth of bio-politics:" Michel Foucault's Lecture at the Collège de France on Neo-liberal Governmentality. *Economy and Society*, 30(2), 190–207.

Loyd, J., Burridge, A., and Mitchelson, M. 2009. Thinking (and Moving) Beyond Walls and Cages: Bridging Immigrant Justice and Anti-prison Organizing in the United States. *Social Justice*, 36(2 (116)), 85–103.

Loyd, J., Mitchelson, M., and Burridge, A. 2012. *Beyond Walls and Cages: Prisons, Borders, and Global Crisis*. Athens: University of Georgia Press.

Lynch, M. 2001. From the Punitive City to the Gated Community: Security and Segregation across the Social and Penal Landscape. *University of Miami Law Review*, 56, 89–112.

Mair, G. and May, C. 1997. *Offenders on Probation*. London: Home Office.

May, C. 1999. *Explaining Reconviction Following a Community Sentence: The Role of Social Factors*. London: Home Office Research Study 192.

May, D.C. and Wood, P.B. 2005. What Influences Offenders' Willingness to Serve Alternative Sanctions? *The Prison Journal*, 85(2), 145–67.

McClintock, A. 2009. Paranoid Empire: Specters from Guantánamo and Abu Ghraib. *Small Axe*, 13(1), 50–74.

McDowell, L. 1997. Introduction: Homeplace. *Undoing Place: A Geographical Reader*, edited by L. McDowell. London: Arnold.

McGarrell, E.F. and Hipple, N.K. 2007. Family Group Conferencing and Re-offending Among First-time Juvenile Offenders: The Indianapolis Experiment. *Justice Quarterly*, 24(2), 221–46.

Millay, J.R. 1991. From Asylum to Penitentiary: the Social Impact of Eastern Oregon Correctional Institution Upon Pendleton. *Humboldt Journal of Social Relations*, 17(1&2), 171–95.

Miller, D. 1998. Why Some Things Matter. *Material Cultures: Why Some Things Matter*, edited by D. Miller. London: University College London.

Miller, D. 2001. *Home Possessions: Material Culture behind Closed Doors*. Oxford: Berg.

Mills, A. and Codd, H. 2008. Prisoners' Families and Offender Management: Mobilizing Social Capital. *Probation Journal*, 55(1), 9–24.

Ministry of Justice (MoJ). 2010. *Compendium of Reoffending Statistics and Analysis: Ministry of Justice Statistics Bulletin*. London: Home Office.

Ministry of Justice (MoJ). 2011. *Adult Re-Convictions: Results From the 2009 Cohort England and Wales*. London: Home Office.

Neumann, A.L. 2000. *California Behind Bars*. [Online]. Available at: http://www.corrections.com/news/article/6817 [accessed: February 22, 2010].

Newman, D. (ed.). 1999. *Boundaries, Territory, and Postmodernity*. London; Portland, OR: Frank Cass.

Ní Laoire, C. 2007. The "green green grass of home"? Return Migration to Rural Ireland. *Journal of Rural Studies*, 23(3), 332–44.

Ní Laoire, C. 2008a. Complicating Host-newcomer Dualisms: Irish Return Migrants as Home-comers or Newcomers? *Translocations*, 4(1), 35–50.

Ní Laoire, C. 2008b. "Settling back"? A Biographical and Life-course Perspective on Ireland's Recent Return Migration. *Irish Geography*, 41(2), 195–210.

Nowicka, M. 2007. Mobile Locations: Construction of Home in a Group of Mobile Transnational Professionals. *Global Networks—a Journal of Transnational Affairs*, 7(1), 69–86.

Office of the Attorney General. 2001. Memorandum: Interviews Regarding International Terrorism to All United States Attorneys and Members of the Anti-Terrorism Task Forces. Unpublished memorandum, November 9.

Ogborn, M. 1995. Discipline, Government and Law: Separate Confinement in the Prisons of England and Wales, 1830—1877. *Transactions of the Institute of British Geographers*, 20(3), 295–311.

Opperman, G. 2012. *Doing Time: Prisons in the 21st Century.* Epsom: Bretwalda.

Paasi, A. 1996. *Territories, Boundaries, and Consciousness: The Changing Geographies of the Finnish-Russian Boundary.* Chichester; New York, NY: J. Wiley and Sons.

Paasi, A. 1999. Boundaries as Social Processes: Territoriality in the World of Flows. *Boundaries, Territory and Postmodernity*, edited by D. Newman. London: Frank Cass.

Pallot, J. 2005. Russia's Penal Peripheries: Space, Place and Penalty in Soviet and Post-soviet Russia. *Transactions of the Institute of British Geographers*, 30(1), 98–112.

Peck, J. 2003. Geography and Public Policy: Mapping the Penal State. *Progress in Human Geography*, 27(2), 222–32.

Pickren, G. 2011. "Factories with fences:" the Prison E-waste Recycling Factory at the Intersection of Green Governance and Labor Market Discipline. Paper to the Annual Meeting of the Association of American Geographers, Seattle, WA, April 15.

Power, E. 2009. Border-processes and Homemaking: Encounters with Possums in Suburban Australian Homes. *Cultural Geographies*, 16(1), 29–54.

Pratt, A. 2005. *Securing Borders: Detention and Deportation in Canada.* Vancouver: UBC Press.

Pratt, J. 2002. *Punishment and Civilization: Penal Tolerance and Intolerance in Modern Society.* London: Sage.

Probyn, E. 1996. *Outside Belongings.* New York, NY; London: Routledge.

Ralph, D. 2009. "Home is where the heart is"? Understandings of "Home" Among Irish-born Return Migrants from the United States. *Irish Studies Review*, 17(2), 183–200.

Ralph, D. and Staeheli, L.A. 2011. Home and Migration: Mobilities, Belongings and Identities. *Geography Compass*, 5(7), 517–30.

Ridgley, J. 2008. Cities of Refuge: Immigration Enforcement, Police, and the Insurgent Genealogies of Citizenship in U.S. Sanctuary Cities. *Urban Geography*, 29(1), 53–77.

Rollenhagen, M. 1999. Wisconsin City Shows Jails Don't Hurt Property Value. [Online]. Available at: http://www.cleveland.com/news/pdnews/metro/ Sunday/ca13hou.ssf [accessed by request].

Sparke, M. 2006. A Neoliberal Nexus: Citizenship, Security and the Future of the Border. *Political Geography*, 30(2), 151–80.

Sparks, R. 2002. Out of the "Digger:" the Warrior's Honour and the Guilty Observer. *Ethnography*, 3(4), 556–81.

Sundberg, J. 2008. "Trash-talk" and the Production of Quotidian Geopolitical Boundaries in the USA-Mexico Borderlands. *Social and Cultural Geography*, 9(8), 871–90.

Tewksbury, R. 2002. Validity and Utility of the Kentucky Sex Offender Registry. *Federal Probation*, 66 (1), 21–6.

Thompson, J. 2000. Critical Citizenship: Boal, Brazil and Theatre in Prisons. *Annual Review of Critical Psychology*, 2, 181–91.

Turner, J. 2012. Criminals with "Community Spirit": Practising Citizenship in the Hidden World of the Prison. *Space and Polity*, 16(3), 321–34.

Turner, J. 2013. Disciplinary Engagements with Prisons, Prisoners and the Penal System. *Geography Compass*, 7(1), 35–45.

US Department of Justice, Bureau of Justice Statistics. 2000. *Sourcebook of Criminal Justice Statistics*. [Online]. Available at: http://www.albany.edu/ sourcebook/ [accessed: August 21, 2013].

Valentine, G. and Longstaff, B. 1998. Doing Porridge: Food and Social Relations in a Male Prison. *Journal of Material Culture*, 3(2), 131–52.

Varley, A. 2008. A Place Like This? Stories of Dementia, Home, and the Self. *Environment and Planning D: Society and Space*, 26(1), 47–67.

Venn, C. 2009. Neoliberal Political Economy, Biopolitics and Colonialism a Transcolonial Genealogy of Inequality. *Theory Culture and Society*, 26(6), 206–33.

Vergara, C.J. 1995. *The New American Ghetto*. New Brunswick, NJ: Rutgers University Press.

Wacquant, L. 2000. The New "Peculiar Institution:" On the Prison as Surrogate Ghetto. *Theoretical Criminology*, 4(3), 377–89.

Wacquant, L. 2001. Deadly Symbiosis: When Ghetto and Prison Meet and Mesh. *Mass Imprisonment in the United States*, edited by D. Garland. London: Sage.

Wacquant, L. 2009. The Body, the Ghetto and the Penal State. *Qualitative Sociology*, 32(1), 101–29.

Walter, B. 2004. Irish Women in the Diaspora: Exclusions and Inclusions. *Womens Studies International Forum*, 27(4), 369–84.

Walter, B. 2006. English/Irish Hybridity: Second-generation Diasporic Identities. *International Journal of Diversity in Organisations, Communities and Nations*, 5, 17–24.

Waltz, K. 1999. *Globalization and Governance*. [Online]. Available at: http:// www.mtholyoke.edu/acad/intrel/walglob.htm [accessed: December 15, 2008].

Webster, R., Hedderman, C., Turnbull, P., and May, T. 2001. *Building Bridges to Employment for Prisoners. Home Office Research Study 226.* London: HMSO.

Yau, N. 2007. Celtic Tiger, Hidden Dragon: Exploring Identity among Second Generation Chinese in Ireland. *Translocations*, 2(1), 48–69.

Young, I.M. 1997. *Intersecting Voices: Dilemmas of Gender, Political Philosophy and Policy.* Princeton, NJ: Princeton University Press.

Young, I.M. 2011. *Justice and the Politics of Difference.* Princeton, NJ: Princeton University Press.

Chapter 12

Conclusion

Corey Johnson

Borders in Everyday Life

Not too far from where I live, but very far from an international land border, Kiawah Island, South Carolina, is part of our bordered world. Kiawah just happens to be my most recent encounter with bordering in everyday life, and at the risk of trivializing the significant trends explored in this edited volume, I open with this anecdote to illustrate one modern face of bordering. A barrier island near Charleston, Kiawah's beautiful beaches and world class golf courses attract a particular class of privileged residents and visitors: mostly white, arriving at the airport and shuttled by van or private car to the island, and capable of living on the self-contained enclave for weeks or more oblivious to the cares of the outside. I went with some friends to find a beach, not knowing that Kiawah, which prides itself on having hosted the 2012 Professional Golf Association Championship and being listed on any number of "best beach" rankings, does not throw open its gates and lay out the welcome mat for just any itinerant beach comber or nature lover. Cars are routed through two lanes of entering traffic, one for residents and one for visitors, and both lanes are strictly controlled by a private security force. After visually assessing the occupants of the car, the "officer" asks your destination and whether you have registered to visit the island—to these authorities the practical equivalent a passport. A reply of "no" quickly earns you a request to turn around and head to the public beaches elsewhere. We asked if it would be possible to visit one of the hotels for lunch, which earned us a yellow permit, a short-term visa on the exclusive island. "Half mile ahead on the right, please," the security officer told us, and we proceeded believing erroneously that we had gained unfettered access to Kiawah. Our yellow permit, however, had marked us as visitors worthy of little else but suspicion, and it kept us from passing subsequent border crossings set up on the island. No interaction with the state occurred during our brief visit, but it had the trappings of a border encounter even if the stakes of this particular encounter were exceedingly low.

Has the once exceptional border encounter become commonplace? The contributors to this volume have collectively sharpened our focus on bordering practices in everyday life in a number of ways. What we aimed to do with this set of interventions was to complicate the connection between borders and the sovereign state by identifying individuals and organizations that engage in border work at a range of scales and places. In the example above, bordering is distilled to what political philosopher Carl Schmitt defined as the essence of the political: the process of differentiating "friend" from "foe" (*Freund-Feind-Unterscheidung*). As recent interest by Anglo-American geography in Schmitt attests, these processes play out in and through space (Elden 2010, Barnes and Minca 2012, Meyer, Schetter, and Prinz 2012, see also Schmitt 1996, Joseph and Rothfuss, this volume).

But this volume does not propose a rigid theoretical framework for understanding the why and where of borders. Instead, we sought to assemble a range of theoretically informed, empirically rich perspectives on a very simple question: Who borders and how? What we received in return exceeded our expectations. The themes that have emerged are numerous, multifaceted, at times troubling, and as we did expect, ultimately largely unresolved. Rather than wrap up all of the loose ends, then, by way of conclusion, we set forth five themes of mobility and control, the vernacular border, border encounters, contested and constructed borders, and the collaboration between state and non-state border workers that connect the chapters in the preceding volume. We hope these themes will offer some possible directions for future work in everyday bordering.

Five Themes in Everyday Bordering

Mobility and Control

Both at the borderline itself as well as on the "inside," several contributors to this book underscore the securitization of all types of mobility in space, alluding to what Didier Bigo has described in his work as "liberal governmentality." This paradigm of modern border security boils down to reframing "freedom" as not being stopped, while security is about impeding mobility for particular subjects—creating for particular subjects under particular circumstances the "governmentality of unease" (Bigo 2011). However, this book also challenges and complements Bigo's focus on borders being everywhere. Importantly, it is not just the state, broadly conceived, that seeks to control mobilities, but also a range of non-state actors. These processes play out in multiple time-spaces (see also Parker and Vaughan-Williams 2009).

The most familiar face of borders as a form of social control happens at the borderline between sovereign states, but the hypermobility that characterizes an increasingly interconnected, globalized world has led to, in the words of Coleman and Stuesse, an "implosion" of bordering not just to sites but to new realms of daily life. Yakubu Joseph and Rainer Rothfuss touch on a similar theme in their chapter, namely the blurring of the distinction between "national security," for which a military was traditionally responsible, and personal security, traditionally the realm of the police. Indeed, at the heart of the Westphalian territorial order was a distinction between the security and defense of the sovereign territory against external threats on the one hand, and the internal policing on the other (Hayes 2009, Hörnqvist 2004). As Judith Miggelbrink shows in her chapter on the EU, and then again in Joseph and Rothfuss's examination of Jos, Nigeria, the merger of biopolitical and territorial forms of enforcement and that merger's impact on social reproduction are neither unique to American immigration policing, nor are they constrained to the scale of nation-states. Indeed, if mobility and control are central themes to modern bordering, then Jen Turner's chapter on prisons provides another useful example of how carceral geographies create boundaries around what is accepted by the state and what is not.

The Vernacular Border

While mobility under the rubric of security will undoubtedly remain central to the study of borders, the contributions to this volume point to a need to push engagement with borders into the everyday ways that humans parcel and experience space. Cooper, Perkins, and Rumford call this emerging agenda the "vernacularization of borders." Borders in everyday life are not always threats to everyone, but also present opportunities for collaboration, resistance, and artistic expression not only across borders, but across scales as well. Sakaguchi Kyōhei's performative bordering in response to the Japanese government's response to 3/11 and the various artists involved in the Mercosul Biennial described by Anne-Laure Amilhat Szary provide examples of the possibilities for citizen action in space. In these examples, borderwork is occurring in ways that rupture, or at least present alternatives to previous orders and how humans understand them. The vernacular border, one that is controlled and produced by citizen actors through language, media, and art, defies what John Agnew called the "territorial trap" at the grassroots level (Agnew 1994) and opens a host of possibilities for citizen action and scholarship.

Border Encounters

If the vernacular border offers opportunities for pushing out the agenda of border studies, we are also reminded throughout the book of our persistently bordered world (Newman 2006). For those who live close to a sovereign state border, the border in everyday life is a reality to challenge, to adapt to, and to overcome, but it is still a reality. Emma Norman shows in Chapter 4 the challenges in mobilizing citizen actors across jurisdictional divides toward a common goal, in her case the mitigation of environmental degradation in Boundary Bay. The theme of encounters between nature, human attempts to manage nature, and the superimposed boundaries that were created with little heed to nature, has been an important avenue of inquiry in political geography and allied disciplines, and there is little doubt it will continue to be.

If in simple terms Emma Norman reminds us that "nature matters" to the border in everyday life, then Kenneth Madsen stresses in Chapter 5 the role that culture plays in the political reality of a bordered world. How humans perceive and internalize territory and belonging differs between cultural groups, but Chapter 5 also reminds us that culture is not a static fixed entity. Similarly, in Vanessa Lamb's chapter on the Hatgyi dam we see a transboundary ethnic minority group, in this case the Karen, that is both inordinately impacted by the state bordering practices and simultaneously not consulted. In each of these three chapters, the colonial border becomes embedded in human–environment relations and the impacted people must negotiate the existence of borders.

Contested and Constructed Borders

Another theme that emerges from this volume involves the ways in which borders are contested and constructed in everyday life. There is no doubt that many borders have been transformed since the early 1980s, which also happens to be when I had my first encounter with the US–Mexico border near dusty Lukeville, Arizona. The most excitement I sensed from my grandparents upon re-entering the US was from the prospect of having to surrender the excess tequila they had stashed somewhere in their Minnie Winnie motorhome. As important work by Jones (2012), Nevins (2010), Wright (2011), and others shows, of course, the contemporary reality is of a heavily militarized border zone where human casualties are the norm, and this particular border bears little resemblance to the one of decades past. In this volume, Jones illustrates how the narrative about borders is shaped by the media, but it would be a mistake simply to dismiss "Border Wars" as a particularly banal form of American "bordertainment." In spite of the creative liberties taken by the show's producers, the plausibility of constructing the Border Patrol and those who cross the US–Mexico border as a crucial nexus in the war against the Other is fed by the reality of a 350 percent increase in the US Border Patrol's budget from 2000–2010 and the attendant new military and surveillance hardware, as well as human capital, that has been placed at the border over the last decade or two.

The materiality of borders meets the discursive in other chapters as well, such as in Lamb's analysis of "discursive governance of the political border"—border talk—or in artistic representations of South America's geo-scapes as impacted by colonization and neoliberalization. Throughout much of the book, examples of borders being constructed and performed in particular ways by state and non-state actors are juxtaposed with contestation of the border, as in the example of the Tohono O'odham intersecting at "The Gate" (Chapter 5) or the mother talking to her son across waters of the Tisza River in the Documentary "The Bridge" (Chapter 7).

Collaboration between State and Non-State Border Workers

Although part of the framing for this volume involved identifying how both state and non-state actors make the border in everyday life, it is also clear for the preceding chapters that the border between those two categories is not neatly drawn. This echoes work by other geographers in critiquing strict state–non-state and public vs. private divides (e.g. Bulkeley and Schroeder 2012). Part of the challenge in border studies moving forward is in identifying how and where the state is in cahoots with non-state actors. The private police at Kiawah Island only can exist with the blessing of and active collaboration with the state. The same is true with the "Border Wars" bordertainment show: celebrating the heroism of the Border Patrol is only possible because the government allowed the crew to accompany them on their various missions. Indeed, every chapter of this book pointed to the expansion of "the state" into a range of different locations and practices that had not previously been thought of as the places where borders are materialized. This expansion of the ordering practices of the state are carried out with, at the minimum, the acquiescence of the population and often through the active participation of individuals and organizations that want to make borders in their everyday lives.

Coda

As the contributors to this volume show in their work, border studies has become decentered from a singular focus on sovereign states and their borderlines (Johnson et al. 2011). Thus the question we began with: who borders and how? This book answered that question by demonstrating that state and non-state individuals are making their own ideas of the border at the border line, official border work is being expanded into the interior of states, non-state individuals are creating boundaries to access away from the borderline, and the idea of what the border signifies is contested and produced by both state and non-state actors, often in collaboration with each other. New technologies and non-traditional border workers are expanding the reach of the sorting and dividing practices of the state into the everyday lives of the population. As a result borders are materializing in new sites, but they are not designed to ensnare everyone. Borders are not everywhere, but they are in a lot more places and are more effective at locating the people they are intended to find.

So where to go from here? We hope that the contributions to this book, individually and collectively, provide if not a practical guide then at least a range of possibilities for examining how borders are made in everyday life.

References

Agnew, J.A. 1994. The Territorial Trap: the Geographical Assumptions of International Relations Theory. *Review of International Political Economy*, 1(1), 53–80.

Barnes, T.J. and Minca, C. 2012. Nazi Spatial Theory: the Dark Geographies of Carl Schmitt and Walter Christaller. *Annals of the Association of American Geographers*, 103(3), 669–87.

Bigo, D. 2011. Freedom and Speed in Enlarged Borderzones. *The Contested Politics of Mobility: Borderzones and Irregularity*, edited by V. Squire. New York: Routledge.

Bulkeley, H. and Schroeder, H. 2012. Beyond State/Non-State Divides: Global Cities and the Governing of Climate Change. *European Journal of International Relations*, 18(4), 743–66.

Elden, S. 2010. Reading Schmitt Geopolitically: Nomos, Territory and Großraum. *Radical Philosophy*, 161, 18–26.

Hayes, B. 2009. NeoConOpticon: The EU Security-Industrial Complex. London: Statewatch and Transnational Institute, Document.

Hörnqvist, M. 2004. The Birth of Public Order Policy. *Race & Class*, 46(1), 30–52.

Johnson, C., et al. 2011. Interventions on Rethinking "the Border" in Border Studies. *Political Geography*, 30(2), 61–9.

Jones, R. 2012. *Border Walls: Security and the War on Terror in the United States, India and Israel*. London: Zed Books.

Meyer, R., Schetter, C., and Prinz, J. 2012. Spatial Contestation?—The Theological Foundations of Carl Schmitt's Spatial Thought. *Geoforum*, 43(4), 687–96.

Nevins, J. 2010. *Operation Gatekeeper and Beyond: The War on Illegals and the Remaking of the U.S.–Mexico Boundary*, 2nd Ed. New York: Routledge.

Newman, D. 2006. The Lines That Continue to Separate Us: Borders in Our "Borderless" World. *Progress in Human Geography*, 30(2), 143–61.

Parker, N. and Vaughan-Williams, N. 2009. Lines in the Sand? Towards an Agenda for Critical Border Studies. *Geopolitics*, 14(3), 582–7.

Schmitt, C. 1996. *The Concept of the Political*. Chicago: University of Chicago Press.

Wright, M.W. 2011. Necropolitics, Narcopolitics, and Femicide: Gendered Violence on the Mexico–U.S. Border. *Signs*, 36(3), 707–31.

Index